# PHOTOVOLTAICS IN BUILDINGS

# PHOTOVOLTAICS IN BUILDINGS

## A Design Handbook for Architects and Engineers

*International Energy Agency, Paris, France*
*Solar Heating & Cooling Programme, Task 16*
*Operating Agent: Heribert Schmidt*


**Principal Editors:**

**Friedrich Sick**
**Thomas Erge**
Fraunhofer Institute for Solar Energy Systems ISE
Freiburg, Germany

Published by James & James (Science Publishers) Ltd, Waterside House,
47 Kentish Town Road, London NW1 8NZ, UK



A catalogue record for this book is available from the British Library

ISBN 1 873936 59 1

Printed in the UK by Bell & Bain Ltd, Glasgow



**Disclaimer**

This book was prepared as an account of work done within Task 16 *Photovoltaics in Buildings* of the IEA Solar Heating and Cooling Programme. Neither the International Energy Agency, nor any of their employees, nor any of their contractors, subcontractors or their employees, makes any warranty, expressed or implied, or assumes any legal liability or responsibility for the accuracy and completeness of any information, apparatus, product or process disclosed, or represents that its use would not infringe privately owned rights

# Table of Contents

**Page**

**Page**

# Preface

The modular structure of photovoltaic systems makes it possible to convert solar energy into electricity over a wide power range from milliwatts to megawatts directly at the place of use. This environmentally acceptable way of generating electricity offers even more advantages when it is applied to buildings. In this case, the building itself acts as a support structure for the modules, and at the same time, PV modules can be an integral part of the building e. g. as a weatherproof roof or facade element or as a shading device. With these advantages, it can be foreseen that PV will become a common component in newly erected as well as retrofitted buildings.

Recognizing this development, the Executive Committee of the Solar Heating and Cooling Programme (SHCP) under the chair of Gerhard Schriber (Switzerland) created the first working group within the International Energy Agency (IEA) dealing with photovoltaics: Task 16 "Photovoltaics in Buildings" was initiated.

Under the leadership of Prof. Jürgen Schmid (Germany), experts from 12 nations started this collaboration in 1990. In 1992, Dr. Heribert Schmidt (Germany) took over the Operating Agent´s duties and managed the Task until its end in 1996.

Task 16 has assessed techniques for maximizing the solar share in total energy concepts as well as optimizing the economics. For system optimization, all energetic aspects such as lighting, heating and cooling, or hot water production have been taken into account.

Results of these findings have been incorporated into Task 16 demonstration projects which were implemented in most of the Task 16 member countries. Both residential and commercial buildings, and grid-connected and stand-alone-systems, are included.

The work was performed in four Subtasks:

In Subtask A (System Design and Engineering, co-ordinating country Finland, Kimmo Peippo) participants produced working documents on existing PV systems, components, energy-efficient electric appliances and lighting equipment, safety issues and national regulations, codes and pricing practices for electricity generation, based on responses to questionnaires. Drawing on this information, recommendations and guidelines for energy concepts, utility interface issues and monitoring procedures were made.

Various methods of integrating PV modules into the building structure have been investigated and tested in Subtask B (Building Integration, Switzerland, Peter Toggweiler). Both architectural and engineering aspects have been taken into consideration. International workshops, an architectural ideas competition under the guidance of The Netherlands, design supporting tools, and the Demosite are amongst the highlights of the Task collaboration. The Task 16 Demonstration Site 'Photovoltaic Buildings Elements' (Demosite) at EPFL, Lausanne, Switzerland, exhibits the best integration methods to architects, engineers and the public.

Using the results of the preparatory work in Subtasks A and B, PV buildings have been designed, constructed and monitored under the leadership of Tony Schoen, The Netherlands, in Subtask C (PV Demonstration Buildings). Most of the Task 16 participating countries have built one or more demonstration buildings. The data from these buildings will be made available to the public.

This design handbook for PV project planners and engineers has been compiled by Subtask D (Technology Communication, Germany, Thomas Erge) and summarizes the information gained in Subtasks A and B and the PV demonstration buildings. To further disseminate the Task 16 results, national workshops, as well as an international symposium with published proceedings, have been organized.

The work proceeded at a very high level thanks to the excellent co-operation of all the experts involved, for which I am very grateful. In particular, my heartfelt gratitude is extended to Prof. Jürgen Schmid, who brought the team together and determined the structure and goals of Task 16. Further thanks are due to Tony Schoen (The Netherlands), Steven Strong (USA) and Peter Toggweiler (Switzerland) for the organization and running of special workshops and conferences.

I very much appreciate the help of Friedrich Sick and Thomas Erge (Germany), who compiled and edited this book, and Ron Laplace (Canada) and James & James (Science Publishers) Ltd. (UK) for proof-reading.

Last but not least, the work could not have been undertaken without the support of the SHCP Executive Committee, and I thank the members sincerely for their cooperation and advice.

Freiburg, March 1996

Dr. Heribert Schmidt
Operating Agent IEA Task 16

## Section A

# GENERAL

## Principal Contributors

**Chapter 1:** **Why Photovoltaics in Buildings?**

Kimmo Peippo (Helsinki University, Finland)
Peter Lund (Helsinki University, Finland)

**Chapter 2:** **The Solar Resource**

Friedrich Sick (Fraunhofer ISE, Germany)

**Chapter 3:** **The Photovoltaic Principle**

Mats Andersson (Catella Generics AB, Sweden)
Jyrki Leppänen (Neste, Finland)

**Chapter 4:** **Types of Photovoltaic Systems**

Friedrich Sick (Fraunhofer ISE, Germany)

Chapter 1

# Why Photovoltaics in Buildings?

## 1.1 Beyond energy conscious design

The potential threat of global climate change, increasing energy demand of the developing world, and inevitably, although not rapidly, diminishing fossil fuel resources have made sustainable energy supply a planetary issue that has to be addressed by literally every sector of human life. At the same time buildings continue to play a significant role in the global energy balance. Typically they account for some 20-30% of the total primary energy requirements of industrialized countries. With increasing awareness of the ecological consequences of energy consumption, the need for energy and environment conscious building design has become more and more pressing.

The building designer already has a number of sustainable technologies to choose from: premium thermal insulation, advanced heating, ventilation and air conditioning (HVAC) equipment, passive solar architecture featuring climate conscious building orientation and advanced glazing and daylighting options; active solar thermal technologies for space heating and domestic hot water; and energy efficient lighting and appliances. All these measures can and already have significantly reduced especially the thermal energy requirements of buildings. This in turn has increased the share of electricity in the energy balance of the building sector.

Until recently it was not feasible to go beyond the energy conscious building design from merely saving to actually producing high value energy and sharing it with the whole of society. But now a new technology, photovoltaics, has emerged as a viable option. Photovoltaics generate electricity from the renewable resource of sunlight and can be installed on or at the actual building, giving a new dimension to energy conscious design.

## 1.2 The photovoltaic option

Photovoltaic (PV) or solar electric modules are solid state devices that convert solar radiation directly into electricity with no moving parts, requiring no fuel, and creating virtually no pollutants over their life cycle. During four decades of photovoltaic activity the devices originally used in space technology have gradually found their way into numerous applications. The state-of-the-art photovoltaic technology today can be characterized as follows:

- PV modules are technically well proven with an expected service time of at least 30 years.
- PV systems have successfully been used in thousands of small and large applications.
- PV is a modular technology and can be employed for power generation from milliwatt to megawatt facilitating dispersed power generation in contrast to large central stations.
- PV electricity is a viable and cost-effective option in many remote site applications where the cost of grid extension or maintenance of conventional power supply systems would be prohibitive.
- PV technology is universal: the PV modules feature a 'linear' response to solar radiation and therefore may be mass produced and shipped world-wide.

Although photovoltaics has the technical potential of becoming a major clean energy source of the future, it is not yet economically competitive in bulk power generation. Instead, it finds its practical applications in smaller scale innovative 'niche' markets, like consumer products, remote telecommunication stations, and off-the-grid dwellings. However, due to rapid technological improvements and the pronounced need for sustainable energy solutions, PV in buildings, also connected to the utility grid, now shows promise of becoming more than just another niche market.

## 1.3 Combining technology and architecture

Traditionally, PV modules or PV arrays have been mounted on special support structures. However, they can also be mounted on buildings, or even be made an integral part of the building envelope thus creating a natural on-site link between the supply and demand of electricity. Through the use of photovoltaics the consumption of power plant based electricity may be significantly reduced. The buildings may even be turned into small distributed net electricity producers and, as such, offer increasing benefits to all.

From an architectural, technical and financial point of view, PV in buildings today

- does not require any extra land area and can be utilized also in densely populated areas,
- does not require any additional infrastructure installations,
- can provide electricity during peak times and thus reduce the utility's peak delivery requirements,
- may reduce transmission and distribution losses,
- may cover all or a significant part of the electricity consumption of the corresponding building,
- may replace conventional building materials and thus serve a dual role which enhances pay back considerations,
- can provide an improved aesthetic appearance in an innovative way,
- can be integrated with the maintenance, control and operation of the other installations and systems in the building,
- can provide reduced planning costs.

Once put in the building context, photovoltaics should not be viewed only from the energy production point of view. Because of the physical characteristics of the PV module itself, these components can be regarded as multifunctional building elements that provide both shelter and power.

Being a mixture of technology, architecture and social behaviour, PV in buildings eludes unambiguous evaluation of its cost-effectiveness and market potential. To a large extent, the value of the concept remains to be assessed on a case by case basis given the economical, technological, architectural, social and institutional boundaries of the project under consideration.

## 1.4 Can PV in buildings make a difference?

Photovoltaics harnesses solar energy, an immense resource that, if fully utilized, could exceed the current energy demand of mankind. But can PV in buildings, today a marginal technology, grow to be more than an exotic option for those who can afford it?
Photovoltaics can be integrated on virtually every conceivable structure from bus shelters to high rise office buildings or even turned into landscaping elements. Although the exact analysis of the potential of PV in buildings calls for careful assessment of several factors including solar availability on building surfaces, institutional restrictions and electric grid stability, it

is easy to become convinced of the large potential of this technology. Even in climates of only moderate solar radiation, the roof top of a single family dwelling can readily accommodate a PV array large enough for electric self-sufficiency on an annual basis. There, PV can certainly make the difference. But PV in buildings can prove to be more than that.

For the vast physical potential of the solar resource and photovoltaic technology to materialize in cost-effective applications, it is crucial that large enough markets emerge to cut down the price per watt of PV. The decreasing cost of photovoltaics would then in turn create an expanding market of new affordable PV solutions. Today, PV in buildings appears as the most promising of these candidate markets to bridge the way for PV from the scattered small-scale niche applications to a major power generating technology of the twenty-first century. In this opportunity also lies the fundamental difference of PV and other energy efficiency options for buildings. Should PV in buildings really trigger the snow ball effect of reducing cost and expanding the use of photovoltaics, it would raise the impact of this concept from being merely a stimulating play ground for architects and engineers to genuinely making the difference in the global perspective.

## 1.5 A mission for architects and engineers

The photovoltaic community may have great visions of the future, but PV in buildings is already an option for today with numerous successful examples. Building design is an integral process and photovoltaic technology adds to the choices available for the energy conscious designer, as this handbook is about to show. It is up to the designer to weigh the pros and cons of the various technologies in each individual project, and make the choice. In short, photovoltaics is worth considering

- if the building has access to solar radiation,
- if innovative design options are preferred,
- if the building is or will be energy-efficient by design.

Although an inherently elegant concept, photovoltaics in buildings is not turned into appealing architecture and sound engineering without the concerted professional efforts of several disciplines. Only by working closely together, can engineers and architects combine technology and architecture in a way that may revolutionize our understanding of both energy and buildings.

# Chapter 2

# The Solar Resource

## 2.1 Sun and solar constant

The sun is a sphere of intensely hot gaseous matter with a diameter of 1.39 x $10^9$ m and is, on the average, 1.5 x $10^{11}$ m from the earth. This distance compares to about 12,000 times the earth's diameter. The eccentricity of the earth's orbit is such that the distance between the sun and the earth varies by 1.7%. The sun has an effective blackbody temperature of 5777 K. The radiation emitted by the sun and its spatial relationship to the earth result in a nearly fixed intensity of solar radiation outside the earth's atmosphere, often referred to as extraterrestrial radiation. The values for this **solar constant** found in the literature vary slightly due to the measurement techniques or assumptions for necessary estimations. The World Radiation Center (WRC) has adopted a value of **1367 W/m²**, with an uncertainty in the order of 1%.

Compared to fossil fuels, the energy density of the solar radiation is relatively small. The total amount of incident radiation, however, is 6500 times larger than the world's energy demand. Even if only the land-covered part of the earth's surface is considered, the sun could still supply 1900 times our worldwide energy demand.[1]

## 2.2 Available solar radiation and spectral distribution

Solar radiation at normal incidence received at the surface of the earth is subject to two significant phenomena:

- atmospheric scattering by air molecules, water and dust and
- atmospheric absorption by $O_3$, $H_2O$ and $CO_2$.

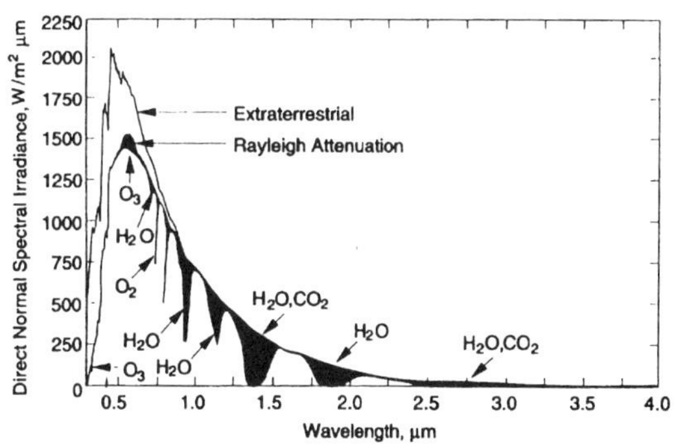


*Figure 2.1 An example of the effects of Rayleigh scattering and atmospheric absorption on the spectral distribution of beam irradiance.* /1/

Figure 2.1 shows the spectral distribution of the extraterrestrial radiation as a function of the wavelength and the effect of scattering and absorption on this distribution for a clear day.

[1] Maybe even more important is the fact that the solar energy incident on our planet is the only continuous source of exergy. Exergy is the valuable part of energy and irreversibly transformed into anergy, the worthless part of energy, during each energy transformation process. When we talk about energy conservation, we really mean exergy conservation, since it is the exergy that we use up. The use of solar radiation to meet our exergy demands means saving exergy stored within thousands of years instead of spending it within a couple of a hundred years.

| | Cloudless blue sky | Misty...cloudy, sun visible as yellowish disc | Cloudy sky |
|---|---|---|---|
| Global radiation | 600...1000 W/m$^2$ | 200...400 W/m$^2$ | 50...150 W/m$^2$ |
| Diffuse part | 10...20% | 20...80% | 80...100% |

*Table 2.1 Irradiance at different weather conditions.*

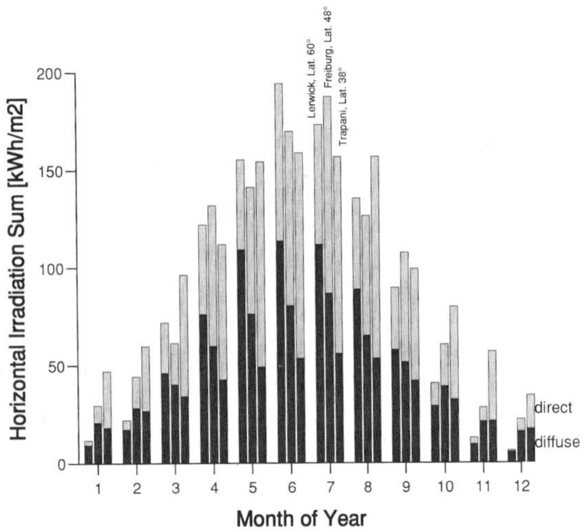


*Figure 2.2 Monthly average daily direct and diffuse irradiation in Lerwick/ Shetland Islands, UK, Freiburg, Germany, and Trapani/Sicily, Italy.*

| Location | Latitude | Annual incident energy [kWh/ m$^2$] |
|---|---|---|
| Sahara | 25°N | 2500 |
| Israel | 33°N | 2000 |
| Trapani, I | 38°N | 1800 |
| Freiburg, D | 48°N | 1100 |
| Helsinki, FIN | 60°N | 950 |
| Lerwick, UK | 60°N | 775 |

*Table 2.2 Annual incident solar energy at several locations on a horizontal surface.*

The ratio of the available global radiation on the horizontal surface and the extraterrestrial radiation for the location is called clearness index. The clearness index hardly exceeds 0.75 on very clear days. The global radiation is made of two parts: the direct radiation from the sun itself and the diffuse radiation from the sky (without the sun).

Table 2.1 gives a rough indication of the relation between weather condition, global radiation and the percentage of diffuse radiation.

In many parts of the world, for example in Central and Northern Europe, the diffuse radiation plays an important role for solar energy conversion: here the diffuse part of the global radiation energy amounts to between 40% (summer) and 80% (winter). Figure 2.2 gives examples for different latitudes in Europe. The annually available radiant energy depends on the geographic location and meteorological conditions. Seasonal changes are due to the tilted axis of the earth on its orbit. Table 2.2 shows that the annually available global radiation may vary by a factor of more than 2.5.

Figure 2.3 illustrates well that seasonal changes have a larger effect on the available radiation at higher latitudes (= degrees North or South from the equator). The images of the globe are taken with the sun's view direction towards the Earth in summer (top) and winter (bottom). While Central Europe is highly 'visible' by the sun during summer noon, it is hardly recognizable during winter. On the other hand, the entire African continent, as an example for low latitudes, is highly exposed to the sun during the whole year.

Figure 2.4 shows how the intensity of the solar radiation on a flat surface is higher when it is tilted towards the sun. The maximum intensity occurs when the flat surface is perpendicular to the sun's rays. Two-axis tracking of receivers may thus maximize the energy gain at the expense of technical complexity. For fixed receiver surfaces, the energy gain is a function of the slope angle (0°: horizontal, 90°: vertical) and the azimuth angle (0°: South, -90°: East, +90°: West, 180°: North). The distribution of the annual incident energy on a tilted surface as a function of slope and azimuth in Central Europe is shown in Figure 2.5. One can observe that there is quite a large region within 90% of the maximum. This gives some freedom in choosing acceptable surfaces for collection of solar energy, which is an important issue for the integration of photovoltaics in building envelopes.

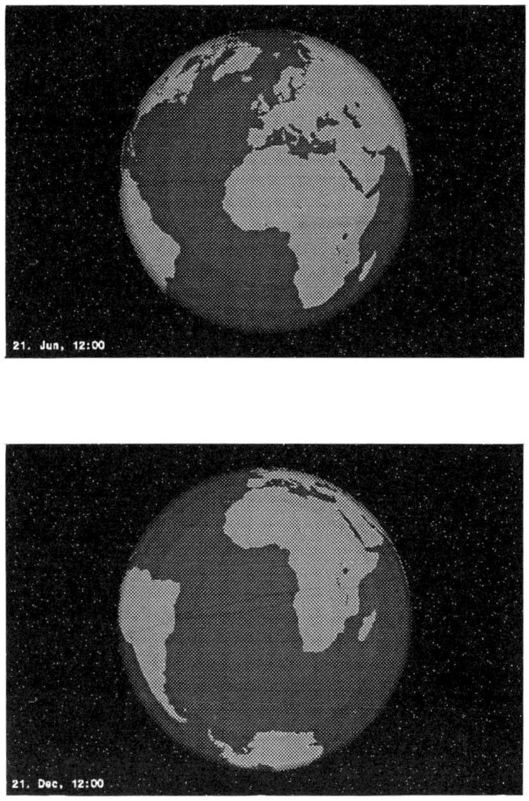


*Figure 2.3 Sun's view of the globe in summer (top) and winter (bottom) around noon in Europe.*

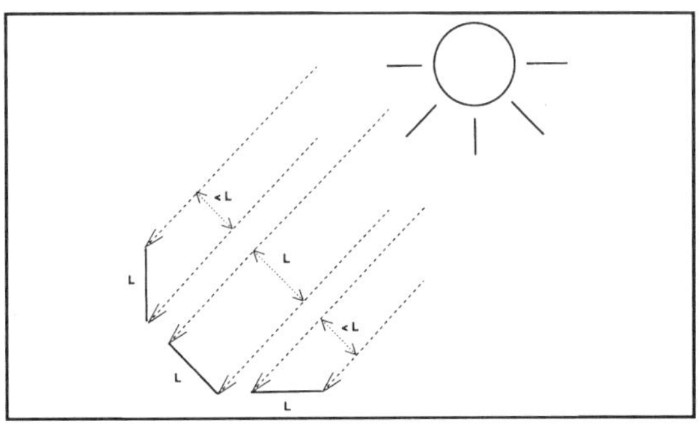


*Figure 2.4 Comparison of a tilted with a horizontal and vertical receiver surface.*

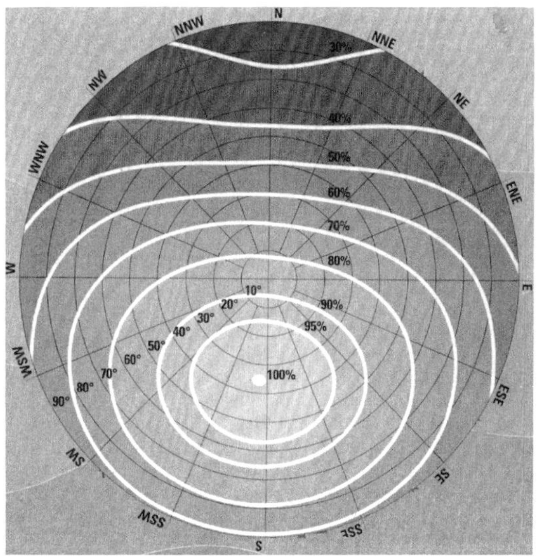


*Figure 2.5 Effect of slope (tilt angle) and azimuth (orientation) on the annual incident energy in Central Europe (© Ecofys).*

/1/ Figure 2.1: John A. Duffie: *Solar Engineering of thermal processes;* Copyright ©1991 by John Wiley & Sons, Inc.; Reprinted by permission of John Wiley & Sons, Inc.

## Chapter 3

# The Photovoltaic Principle

### 3.1 Introduction

The physical phenomenon responsible for converting light into electricity - the photovoltaic effect - was first observed by a French physicist, Edmund Becquerel, in 1839. He noted that a voltage appeared when one of two identical electrodes in a weak conducting solution was illuminated. The photovoltaic effect can be described simply as follows: light, which is a form of energy, enters a photovoltaic (PV) cell and transfers enough energy to cause the freeing of electrons. A built-in potential barrier in the cell acts on these electrons to produce a voltage which can be used to drive a current through an electric circuit.

The first cells were made from selenium during the last century with only 1 - 2% conversion efficiency. Since then, significant research has been done in this field. Quantum mechanics, developed during the 1920s and 1930s, laid the theoretical foundation for our present understanding of PV. However, a major step forward in solar-cell technology was done during the 1940s and early 1950s when a method called the Czochralski method was developed for producing highly pure crystalline silicon. Other important triggers for the PV industry were the space programmes started in the 1950s and also the development of the transistor industry. Transistors and PV cells are made from similar materials, and many of their working principles are determined by the same physical mechanisms.

### 3.2 Cell structure

The basic element in the photovoltaic module is the solar cell which absorbs sunlight and converts it directly into electricity. Figure 3.1 shows the basic structure of a PV cell.

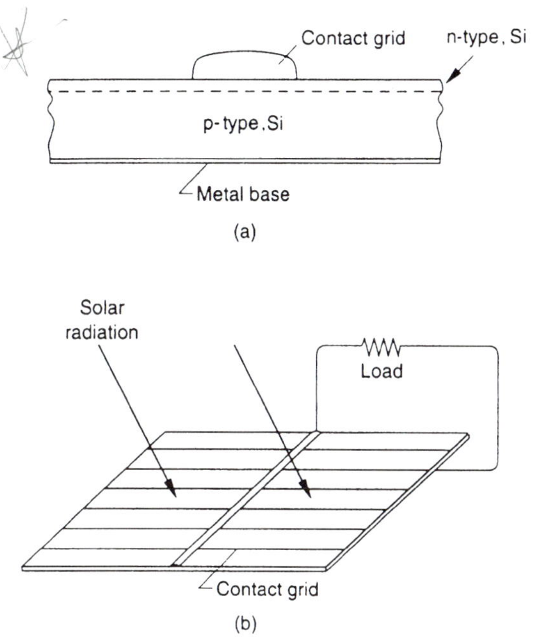


*Figure 3.1 (a) Section of a silicon solar cell. (b) Schematic of a cell, showing top contacts.*

The solar cell consists of a thin piece of semiconductor material, which in most cases is silicon. A semiconductor is an element, whose electrical properties lie between those of conductors and insulators, making it only marginally conductive for electricity.

Through a process called 'doping' a very small amount of impurities is added to the semiconductor, thus creating two different layers called n-type and p-type layers. An n-type material has an increased number of electrons in the conduction band (n=negative) whereas the p-type material has vacancies of electrons (p=positive). Typically, phosphorus is used to create the n-type layer and silicon doped with boron makes the p-type layer. Between these two layers a p-n junction is created which is of great importance for the function of the solar cell.

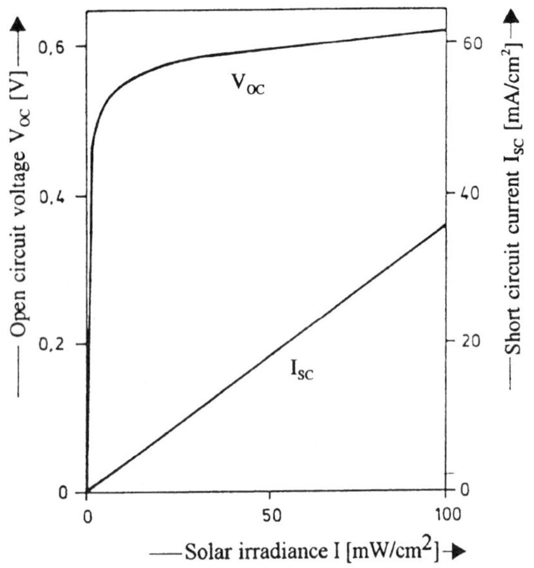


*Figure 3.2 Voltage and current of a silicon solar cell as a function of irradiance.*

The light passes through a 'window layer' which is thin and therefore absorbs only a small fraction of it. The major part of the light is absorbed in the absorber layer where it creates free electrons that can flow through a wire connected to both sides of the cell. In order to do so, a built-in electrical field is needed. This field is formed along the zone or junction between the two layers of n- and p-type silicon. The current produced by the cell is proportional to the amount of incident light (the number of photons entering the cell). Therefore, current increases with the cell area as well as with the light intensity. The voltage, on the other hand, depends on the material used. A silicon cell produces about 0.5 V regardless of cell area. Figure 3.2 illustrates these effects.

## 3.3 Types of photovoltaic cells

### Crystalline solar cells

The most commonly used cell material is silicon (see Figure 3.3). PV cells made of single-crystal silicon (often called monocrystalline cells) are available on the market today with efficiencies close to 20%. Laboratory cells are close to the theoretical efficiency limits of silicon (29%). Polycrystalline silicon is easier to produce and therefore cheaper. It is widely used, since its efficiency is only a little lower than the single-crystal cell efficiency. Gallium arsenide (GaAs) is another single-crystal material suitable for high efficiency solar cells. The cost of this material is considerably higher than silicon which restricts the use of GaAs cells to concentrator and space applications.

### Thin-film solar cells

In order to lower the cost of PV manufacturing, thin-film solar cells are being developed by means of using less material and faster manufacturing processes. The major work on thin films during the last 10 years has been focused on amorphous silicon (a-Si). PV modules made of a-Si are shown in Figure 3.4. The long-term advantage of amorphous as compared to crystalline silicon is the lower need for production energy leading to shorter energy payback time. With the use of small a-Si cells in pocket calculators, a new market, the consumer PV market, was born. The disadvantage of these cells is the relatively low efficiency that has prevented the breakthrough in the production of power in large installations. However, in building applications, a larger module area for the same nominal power due to the lower efficiency may result in a more uniform appearance and thus become advantageous. Although a-Si cells over 10% efficiency are being produced, this initial value is reduced by approximately 30% due to the light-induced instability called Stabler-Wronski effect. Current research focuses on ways to reduce this effect and to increase the efficiency.

Other interesting thin-film materials are Cadmium-Telluride (CdTe) and Copper-Indium-Diselenide ($CuInSe_2$ or CIS). Nowadays, cells made of these materials are produced in laboratories with efficiencies of about 15%. Thin-film crystalline silicon on ceramic substrates is another possible solution being examined today.

In Table 3.1 the most common solar cell materials are summarized.

*Figure 3.3 Crystalline silicon ingot, wafers and PV cells made from these (InterNAPS Ltd.).*

*Figure 3.4 PV modules made of amorphous silicon (a-Si) material (NAPS).*

| Material | Theoretical efficiency | Laboratory cell (1994) | Module (1994) |
|---|---|---|---|
| Crystalline Silicon | 29% | 23% | 15% |
| GaAs | 31% | 25% | |
| **Thin film** | | | |
| Amorphous Si | 27% | 12% | 8% |
| CIS | 27% | 17% | 11% |
| CdTe | 31% | 16% | 10% |

*Table 3.1 Theoretical and practical efficiencies of different types of solar cells.*

## Chapter 4

# Types of Photovoltaic Systems

### 4.1 Introduction

Photovoltaic (PV) systems are of a modular nature. Solar cells can be connected in series or parallel in virtually any number and combination. Therefore, PV systems may be realized in an extraordinarily broad range of power: from milliwatt systems in watches or calculators to megawatt systems for central power production. Building power supply systems are usually in the range of several kilowatts of nominal power.

There are two basically different PV systems: those with a connection to an (available) electric grid and remote or 'stand-alone' systems. While in the first case the grid serves as an ideal storage component and ensures system reliability, the stand-alone systems require a storage battery. This battery serves as a buffer between the fluctuating power generated by the PV cells and the load. In order to ensure continuous power supply, even under extreme conditions, a back-up generator is often also installed. Building-integrated PV systems have an economic advantage over conventional PV generator systems: the PV modules serve multiple purposes. They are part of the building envelope, ideally replacing conventional facade or roof material. Modern commercial building facades often cost as much as a PV facade which means immediate or short-term pay-back for the PV system. Depending on the type of integration, the PV modules may also provide shading or noise protection. Here again, the costs for replaced conventional means for these purposes may be deducted from the initial PV costs.

### 4.2 Grid-connected systems

PV systems may be connected to the public grid. This requires an inverter for the transformation of the PV-generated DC electricity to the grid AC electricity at the level of the grid voltage. National and even regional regulations differ widely with respect to the policy of interconnection requirements and reimbursing for PV-generated electricity fed into the grid. In order to support the production of PV-generated power, some utilities offer a better price for the kWh fed into their grid than they charge for the kWh from the grid. In other locations a one-to-one ratio is applied which means the same kWh-price for both flow directions. The third version is to pay less for the generated PV power fed into the grid than for that sold to the consumer. Comparing the rates, the fixed rates for the power connection have to be considered also. Depending on the kind of tariffs, one or two electricity meters have to be used at the point of utility connection. Figure 4.1 shows a block diagram of a grid-connected PV system suitable for building integration.

In grid-connected applications, photovoltaic systems must compete against the cost of the conventional energy source used to supply the grid. PV systems are particularly cost-effective when the utility load and solar resource profiles are well matched. This is, for example, the case in areas with high air-conditioning loads that have their peaks during the peak sunshine hours of the summer day.

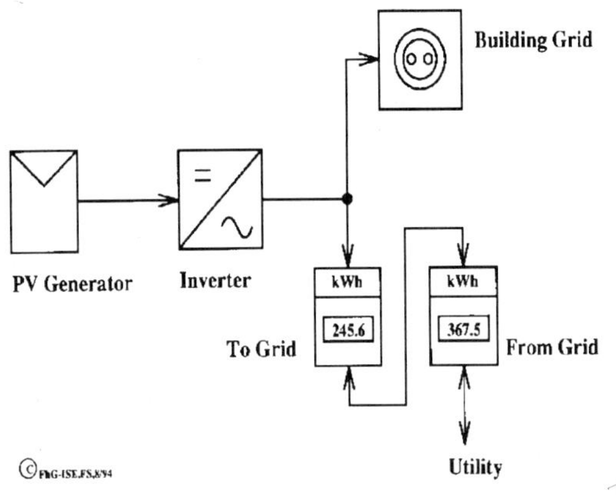


*Figure 4.1* *Principle schematic of a grid-connected PV power system.*

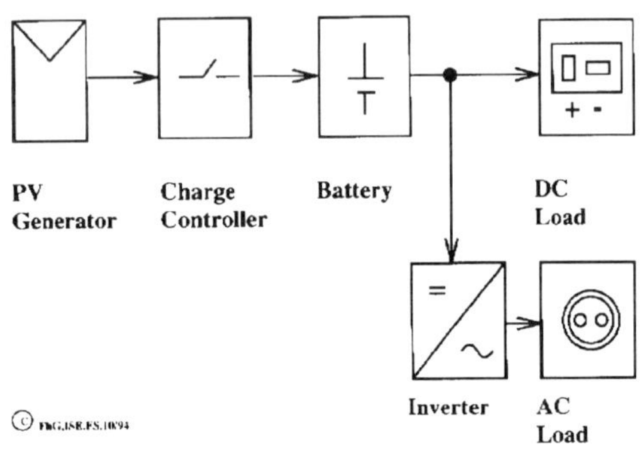


*Figure 4.2* *Principle schematic of a stand-alone PV power system.*

## 4.3 Stand-alone systems

PV systems are most effective at remote sites off the electrical grid, especially in locations where the access is possible by air only, e.g. in alpine regions. Their high reliability and low servicing requirements make them ideally suited for applications at (for parts of the year) unattended sites. The costs for a PV system compete in this case against the cost for a grid-connection or other possible ways of remote energy supply.

As stated above, a storage battery is needed. Excess energy produced during times with no or low loads charges the battery, while at times with no or too low solar radiation the loads are met by discharging it. A charge controller supervises the charge/discharge process in order to ensure a long battery lifetime. As in the grid-connected systems, an inverter, when required, transforms DC to AC electricity. A scheme of such a system is shown in Figure 4.2.

By virtue of the variable nature of the energy source sun, one of the most expensive aspects of a PV power system is the necessity to build in system autonomy. Autonomy is required to provide reliable power during 'worst case' situations, which are usually periods of adverse weather, seasonally low radiation values or unpredicted increased demand for power. The addition of autonomy could be accomplished by oversizing the PV array and greatly enlarging the battery storage bank - generally the two most costly system components. By incorporating the additional battery charging and direct AC load supply capabilities of an engine generator (genset) into the PV system design (as shown in Figure 4.3), the need to build in system autonomy is greatly reduced. These systems are often referred to as hybrid systems. When energy demands cannot be met by the PV portion of the system for any reason, the genset is automatically brought on line to provide the required back-up power. Substantial operating cost savings (compared to a genset system without PV) are achieved through the greatly reduced need for genset operation. An additional benefit of this approach is the added system reliability provided by the incorporation of the back-up energy source.

Hybrid systems may contain more than one renewable power source. Adding a wind turbine to a PV genset system is a common combination in areas with high wind energy potential like coastal or hilly regions. Very often the instantaneously available wind energy is high, while the radiation values are low and vice versa.

## 4.4 Direct use systems

There are applications where the load matches the available radiation exactly. This eliminates the need for any electricity storage and backup. A typical example is the electricity supply for a circulation pump in a thermal collector system.

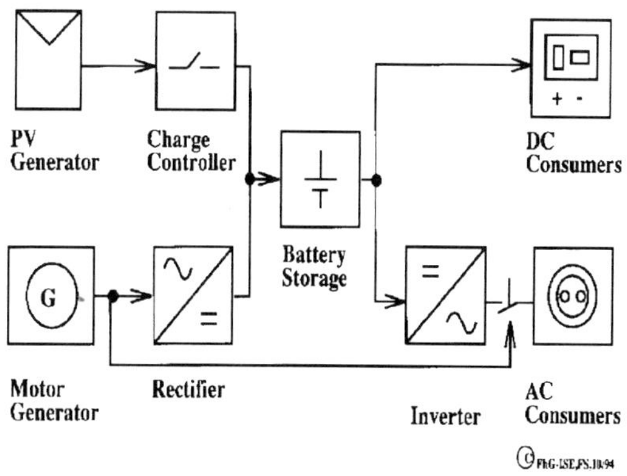


*Figure 4.3 Principle schematic of a hybrid PV power system.*

## Section B

# COMPONENTS

## Principal Contributors

**Chapter 5:** **Photovoltaic Modules**

Mats Andersson (Catella Generics AB, Sweden)

**Chapter 6:** **Photovoltaic Generator**

Hermann Laukamp (Fraunhofer ISE, Germany)

**Chapter 7:** **Energy Storage**

Jyrki Leppänen (Neste, Finland)
Steven Gust (Neste, Finland)
Subrah Donepudi (ESTCO Energy, Canada)

**Chapter 8:** **DC Power Conditioning**

Heribert Schmidt (Fraunhofer ISE, Germany)

**Chapter 9:** **Inverters**

Heinrich Wilk (OKA, Austria)
Erik Wildenbeest (ECN, Netherlands)

**Chapter 10:** **Hybrid Power Systems**

Ron LaPlace (Photron, Canada)

Chapter 5

# Photovoltaic Modules

## 5.1 Introduction

In every building-integrated PV system, the PV module is the basic element of the generator. The number of modules in series will determine the system voltage and the current of the plant can be sized by parallel connection of module strings. The desired output power is the product of system voltage and current.

## 5.2 PV modules

One single silicon solar cell with a surface area of approximately 100 cm$^2$ generates a current of 3 A at a voltage of 0.5 V when exposed to full sunshine. When PV modules first came into terrestrial use, the most common application was to charge 12 V lead-acid batteries requiring a module voltage of 13 to 15 V. Therefore, the typical PV module made of crystalline silicon consists of 30 to 36 cells connected in series with a peak power of approximately 50 W.

A cross-section through a module is shown in Figure 5.1. The module's top layers are transparent. The outermost layer, the cover glass, protects the remaining structure from the environment. It keeps out water, water vapour and gaseous pollutants which could cause corrosion of a cell if allowed to penetrate the module during its long outdoor use. The cover glass is often hardened (tempered) to protect the cell from hail or wind damage. A transparent adhesive holds the glass to the cell. The cell itself is usually covered with an anti-reflective (AR) coating. Some manufacturers etch or texture the cell surface to further reduce the reflection.

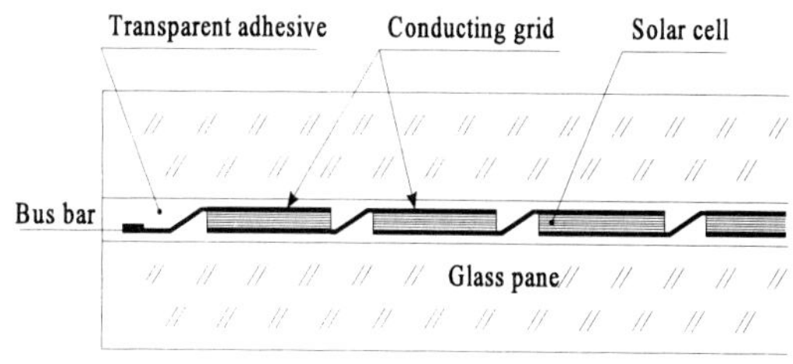


*Figure 5.1 Cross-section through a typical module.*

After the light has passed through the cover glass, the transparent adhesive and the AR coating, it penetrates into the semiconductor material where the electricity is generated. The light-generated current flows out of the cell surface through a metal grid, called the front contact. To reduce resistance losses, it is important that the metal grid is covering parts of the cell surface. On the other hand, blocking a large fraction of the light entering the cell should be avoided. The cell's bottom layer is called the back contact and is a sheet of metal which in connection with the front contact forms a bridge to an external circuit. The module's back side is covered with a layer of Tedlar™ or glass. Often a frame of aluminium or composite material gives the module the needed mechanical stability for mounting it in different ways.

During the last few years, grid-connected systems have been a growing application for PV and systems of several thousands of kW have been built. The system voltages in these plants are sometimes as high as 500 to 1000 V and a large number of modules are connected in series. For these purposes manufacturers have developed large area modules of several square meters with peak power outputs of several hundred watts.

There are a large number of module manufacturers on the market and each company may have 5 to 10 different module types. Therefore, there are a variety of modules to choose from.

*Figure 5.2 Large-area PV module for building integration.*

## 5.3 Modules for buildings

Standard modules are widely used for building applications, especially to retrofit existing buildings. Their frame, however, impedes an easy and elegant integration into roof and facade.

Omitting the frame leads to 'laminates', which can be mounted like glass panes using conventional glazing construction techniques. A specific method of roof integration are 'PV Tiles'. PV Tiles, by their design, can be installed very quickly by electrical lay persons. They combine an old and a new technology and conserve the basic appearance of very common roof types.

The increased interest in PV facades has created new options for customized PV modules. As an example Figure 5.2 shows a large-area PV module for building integration. To allow a larger creativity, module manufacturers are offering several, sometimes even customized, sizes and options to modify the modules' appearance. Chapter 12 covers this topic in more detail.

## 5.4 Definitions, characteristics and performance

In Figure 5.3 characteristic values for a module are shown in an I-V curve. The indicated parameters are explained below. The curve represents the performance at 'Standard Test Conditions' (STC), which is a definition used to compare different modules. STC represent an irradiance of 1000 W/m$^2$ at an Air Mass of AM1.5 and a cell junction temperature of 25°C. AM0 corresponds to the solar spectrum in outer space; at the equator the average spectrum is AM1 and the reference spectrum for STC was defined to be AM1.5 (average spectrum at 45° lattitude).

**$V_{OC}$** is the open circuit voltage of the module, i.e. the voltage of the module when no current is drawn. The $V_{OC}$ is dependent on the cell temperature and decreases with increasing temperature by approximately 0.4%/K for crystalline material. This value is lower for amorphous cells.

**$I_{SC}$** is the short circuit current of the module. Contrary to most other (voltage driven) power sources, a PV module has a short circuit current that is only slightly higher than the operational current. $I_{SC}$ slightly increases with increasing cell temperature by approximately 0.07%/K.

**$P_n$** is the nominal power that the module can produce under STC and it is the value that is given on the manufacturer's plate on the back side of the module. This power value is often referred to as the peak power of the module (**$W_p$**). **$U_n$** and **$I_n$** are the corresponding voltage and current values at this point.

The **Maximum power point (MPP)** is the operating point on a current-voltage (IV) curve where maximum power is produced. For a typical silicon cell this is at about 0.45 V.

The ideal solar cell or module has an I-V curve of rectangular shape (see dashed lines in Figure 5.3). The **fill factor (FF)** indicates the ratio between a real and this ideal cell. Typical values for the FF at STC are between 0.6 and 0.8.

## 5.5 Reading the data sheet

The PV module data sheets must be read with caution because at present there is no uniform way of presenting information. A hypothetical PV module data sheet is provided in Table 5.1.

The example data sheet provides the values at STC. It also notes that the spread of the values is ± 10%. Thus the module power may vary anywhere in the range 44-53 W. Further, in actual operation the module will usually not produce the rated power because the solar cells are sensitive to temperature. For the silicon solar cells, the voltage will be derated by 0.0022 V/K rise in temperature above the STC temperature (the current changes only marginally). In bright sunlight, the module temperature is typically 20 ... 40 K above the ambient temperature. Of course, in cold climates, the module could deliver more than its rated power.

Unless the system includes a maximum power point tracker (see chapter 8.2.1), the system typically will not operate at this point. The designer must determine a typical operating point for the system in question and base the module output calculations on this. In stand-alone systems, the system voltage varies only slightly around the battery voltage. In general, the module peak power point voltage $V_{MPP}$ should exceed the voltage to which the battery is charged by approximately 1.5 V.

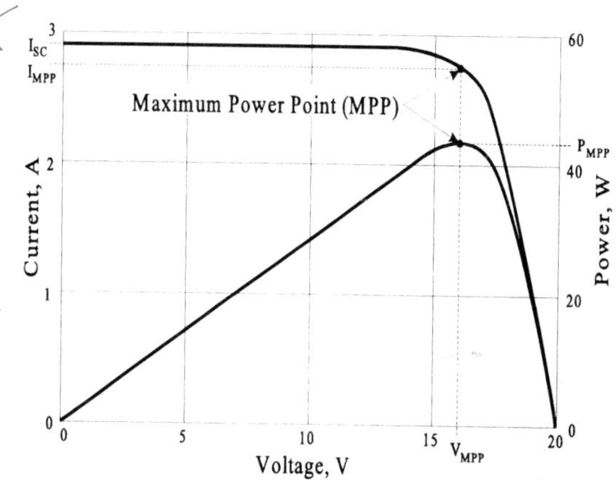


*Figure 5.3 I-V curve and characteristic parameters for a typical module.*

**Module specifications for XYZ48**

**Electrical specifications:**

| | |
|---|---|
| Short circuit current $I_{SC}$ | 3.3 A |
| Open circuit voltage $V_{OC}$ | 21.3 V |
| Current at peak power $I_{MPP}$ | 3.0 A |
| Voltage at peak power $V_{MPP}$ | 16.7 V |
| Maximum power output at 1000 W/m$^2$ and 25 °C | 48.6 W |
| Variation (spread) ± 10% | |
| Voltage decrease with temperature increase | 0.0022 V/K/cell |

**Mechanical specifications:**

| | |
|---|---|
| Front cover | Low iron tempered glass |
| Encapsulant | Ethylene Vinyl Acetate (EVA) |
| Backing | White Tedlar™ |
| Solar cells | 100mm x 100mm square cells, 36 in series |
| Edge sealant | Butyl rubber |
| Frame | Silver anodized structural aluminium |
| Termination | Waterproof junction box |
| Electrical isolation | 3000 VDC 10 μA (TYP) |
| Weight | 6.2 kg |

**Environmental conditions:**

| | |
|---|---|
| Ambient temperature | -40°C to +90°C |
| Wind loading | max. 80 km/h |
| Relative humidity | 0 to 100% |

Thermal shock, hail impact and other environmental conditions as per JPL Block V testing. These modules are covered by the standard ten-year limited warranty on power output. Specifications are subject to change without notice.

*Table 5.1: An example of a typical PV module data sheet.*

## Chapter 6

# Photovoltaic Generator

### 6.1 Introduction

This chapter introduces the fundamental knowledge necessary to successfully install a PV generator. Physical characteristics, radiation influence and shading effects are explained, types of PV arrays and mounting technologies are introduced, hazards and their remedies are stated and a principal block diagram of a PV generator is given.

The reader should keep in mind that although these considerations apply to all PV systems, they focus on 'PV in buildings'. Installed in a facade or a sloped roof, the PV generator will be a factor in the appearance of a building. Thus, it needs special attention with respect to mechanical and electrical as well as aesthetic integration.

### 6.2 Parameters affecting the energy output

A number of parameters affect the possible energy yield of a PV generator (Table 6.1). The most important one is the solar radiation, which is essentially determined by the geographic location and the generator's tilt and orientation.

Further factors to be accounted for include (partial) shading, mismatch of modules in a string, the module operating temperature, resistance of wires and cables, string diodes and soiling.

The effect of generator tilt and orientation on the possible energy yield depends on the ratio of direct to diffuse irradiation. For Central European climate conditions Figure 2.5 shows the relative irradiation on an arbitrarily oriented fixed plane. It is obvious that the exact orientation is not critical. In a wide range of possible orientations more than 95% of the maximum energy is received.

| |
|---|
| Incident radiation |
| Module temperature |
| Partial shading |
| Mismatch of string modules |
| Wire resistances |
| Module soiling |
| String (blocking) diodes |

*Table 6.1: Influences on energy output.*

This statement holds true for an unobstructed PV generator. At locations where soiling, snow, obstacles or distinct daily or seasonal weather patterns occur, these influences have to be taken into account.

Shading is a critical issue. The PV generator performs best if it is homogeneously illuminated. A small shadow from a leaf, an antenna pole, a chimney or an overhead utility line may seriously decrease the available output power. This is due to the fact, that **the cell with the lowest illumination determines the operating current of the whole series string.** This effect is illustrated in Figure 6.1. It is comparable to a water hose, which is pressed tight at one point, preventing the flow of water in the whole hose.

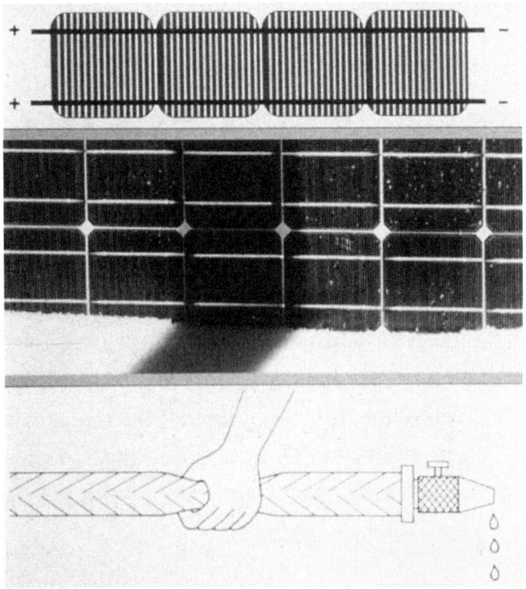

*Figure 6.1 Minor shading can cause a major energy loss.*[1]

Under certain circumstances a partially shaded cell may even be forced into a load mode. This can lead to a thermal destruction of the cell and the respective module. In order to avoid this 'hot spot' effect, 'bypass diodes' are used to provide a second current path diverting the current from the shaded cell.

Amorphous silicon modules, whose cells are long narrow stripes, are less affected by shading than crystalline silicon modules. This is because typical shading objects like trees or street lamp poles usually do not shade a cell over its whole length. Thus in practice the cell current of amorphous silicon cells is only affected by a percentage proportional to the area shadowed. Furthermore, these modules are less susceptive to 'hot spot' development.

A similar effect to partial shading occurs, when modules with different I-V curves are connected in series (module mismatch). The 'weakest' module determines the current through all series-connected modules. Therefore, modules within a series string should be closely matched in order to keep mismatch losses as low as possible.

The temperature influences the modules' efficiency by approximately -0.4%/K. For instance, a 15% efficiency decreases to 14.4% at a temperature increase of 10 K. If feasible, modules should be freely vented. An air gap of 10 cm is sufficient in most cases.

Soiling, i.e. accumulation of dust and dirt may reduce the available generator output. Its effect depends mainly on the source of the dust and the tilt angle of the generator. Dust from nearby industrial complexes, major highways and major railway stations may cause a power reduction up to 10%. Usually dust accumulation and self cleaning reach a steady state after some weeks, if the generator tilt angle is at least 15°. In no case it has been economically attractive to regularly clean the modules. In typical residential areas of 'moderate climate' zones soiling can be neglected.

In areas of heavy snowfall additional considerations should be noted. If a continuous PV output is desired the PV generator should be mounted rather steeply, at least 45°, to allow quick shedding of the snow. A smooth surface eases sliding of the snow. Ideally large, vertically oriented, frameless modules should be used.

## 6.3 Types of arrays and mounting technologies

PV generators on buildings are usually fixed. There are several options for their placement:

- on a sloped roof
  - stand-off
  - integral (modules/tiles);

[1]Pacer program on Photovoltaics; Bundesamt für Konjunkturfragen, Bern, Switzeland.

- at the facade
  - as wall element
  - as protruding shading element;
- on flat roofs
  - stand-off
  - integral.

**'Stand-off'** is a straightforward mounting method well suited for retrofits. Special mounting elements like hooks or mounting tiles are fixed to the roof. A support structure to which the modules are bolted is fixed to the mounting elements. Cable channels collect and protect the string cables, which lead into the building through watertight 'feed-throughs', e.g. modified venting elements. Since the cables are exposed to the outdoor environment, they need to be selected accordingly. The support structure should be designed for at least a 30 years lifetime. Thus, aluminium, stainless steel and glass fibre should be the preferred materials.

**Integral mounting** leads to a nicer appearance and cost savings in new buildings. This method uses the PV generator as the building envelope. The modules replace conventional roof or facade covers. This is accomplished by using frameless modules (so-called laminates) in combination with mounting technologies taken from conservatory or conventional glazing construction. The wiring is usually not exposed to ambient conditions. However, the access to the wiring is more difficult, if thermal insulation is installed in the roof. To remove the air warmed up by the generator efficient venting elements may be included.

A special design of roof-integrated PV modules are **PV tiles**. Prewired tiles can be mounted and connected very quickly and they are accessible from the outside. Several manufacturers offer PV tiles.

**Facades** are an increasingly popular location for PV generators, since they provide multiple purposes for the PV modules. Besides electric power production, PV modules may serve to present corporate identity. Semitransparent modules may serve for daylighting purposes. Installed in front of the facade, modules provide shading for the offices behind. PV facades usually rely on mounting methods for conventional facade elements. Popular mullion/transom constructions (⇒App. VIII) have been modified to allow integral cabling. Also, structural glazing technology has been successfully used for PV facades.

On **flat roofs**, PV generators can be installed using very similar techniques as for PV arrays in the open field. In order to avoid penetration of the roof, 'weight foundations' are often used, which keep the modules down by gravitation.

## 6.4 Block diagram and components of a PV generator

A PV generator comprises a variety of components. These include: modules, fixing material, mounting structure, bypass diodes, blocking diodes, fuses, cables, terminals, overvoltage/lightning protection devices, circuit breakers and junction boxes. Figure 6.2 gives a schematic diagram of a PV generator and its basic design.

**Standard modules** come with about 20 V open circuit voltage ($V_{oc}$) and approximately 3 A nominal current ($I_n$) (at standard test conditions). For higher power ratings the modules are connected in series and/or in parallel. Several modules in series are called a 'string'. In some cases it is necessary to protect the string cables and modules against overcurrent. Fuses are used for general overcurrent protection, while blocking diodes prevent current flow into one string from the rest of the PV generator in case this string does not reach its designed operating voltage for whatever reason. However, considering that PV modules are current sources and that modern cables and appropriate wiring methods make a short circuit or a ground fault extremely unlikely, these protection devices may be omitted in some cases.

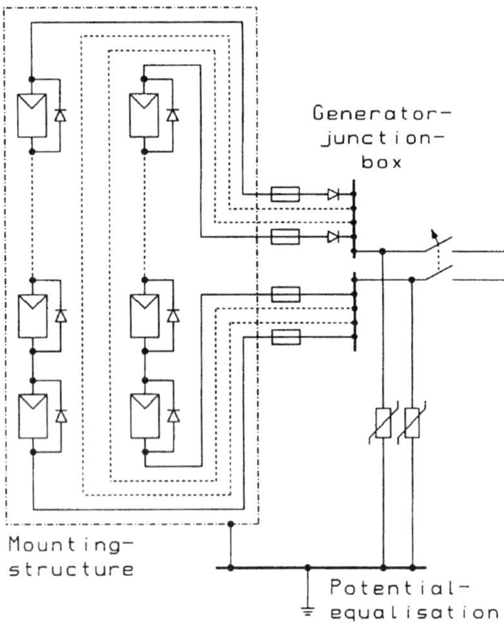


*Figure 6.2 Basic structure of a PV generator.*

In PV strings with open circuit voltages $V_{oc}$ higher than 30 V, '**bypass diodes**' are usually integrated. A bypass diode provides a current path around a module or a part of a module. It protects the bypassed cells in the module, e.g. under partial shading conditions, from operation in a load mode and possible destruction. The need for bypass diodes depends on the system configuration and module specifications.

**Blocking diodes** prevent current flow backwards into a string. This - rather unlikely - operating condition might occur, if ground faults or short circuits happen in a string. It would reduce the generator's power output and at worst could lead to a destruction of cables and modules. However, using modern 'protection class II' modules and 'ground fault proof and short circuit proof' wiring virtually eliminates the occurrence of such a failure.

**Fuses** protect cables from overcurrent. In PV generators they should be used only if a large number of strings is connected in parallel and the generator's short circuit current could exceed the cable's rated current in one string. In many residential systems the intermodule cables can carry currents of several parallel strings without being overloaded. For instance, most systems in the German 1000 Roofs Programme employ 2.5 $mm^2$ cables for string cabling, which are listed for 12 A at 70°C operating temperature when installed in bundles. Thus, under conservative assumptions fuses would be required only in case of more than four strings in parallel assuming standard modules.

**Cables** are usually double-insulated and UV-resistant. They must withstand the elevated temperatures behind the modules. These temperatures can reach 50 K above ambient temperature, if the module back side is covered with thermal insulation material. The size of the cable is determined by the allowable voltage drop along the string at nominal current and thus larger than the nominal operating current of approximately 3 A would require. Using different colours for the + and - connections eases wiring of the junction box.

**Connections** are numerous and very important. A sloppy connection may render a whole string useless or even, in the worst case, cause a fire. Crimp terminations and spring loaded cage clamp terminals are considered most reliable. Plug/receptacle types of connectors are currently being studied, because they offer quick field wiring as well as easy module replacement.

**Overvoltage/lightning protection devices** will keep voltage transients out of the systems. Modern modules are rugged, so they can easily withstand surge voltages up to 6 kV. Electronic components such as bypass or blocking diodes and equipment such as inverters and charge controllers, however, need protection. Therefore, surge arrestors with at least 5 kA peak current ratings are applied at either leg of the PV generator.

**Circuit breakers** between the PV generator and the inverter or charge controller are needed to remove the PV generator's voltage from the main DC line. They must be rated for the generator's nominal short circuit current and open circuit voltage and for DC!

The above-mentioned components are located and electrically connected in one or more **junction boxes** (see Figure 6.3). This box must be suited for the mounting location in terms of IP-protection (☞App. VIII), temperature rating, UV-resistance etc. It should be easily accessible to regularly check the fuses and the overvoltage protection devices and to open the DC circuit breaker(s).

*Figure 6.3 Junction boxes.*

The **mounting structure** holds the modules in place. It must take all mechanical loads, potential wind loads, snow cover, and thermal expansion/contraction with an expected lifetime of at least 20 years. In building applications water tightness is often needed as well. Module mounting and wiring should be simple. The replacement of individual modules should be possible without dismantling the whole PV generator. Advanced mounting structures for PV facades provide easy laying of the string cables, e.g. in integrated ducts.

## 6.5 Hazards and protection

PV modules (and generators) are current sources. There are differences from common electric power sources like the public grid, a motor generator or a battery. This requires some mental reorientation for all those working with either technology. Furthermore PV generators cannot be switched off. As long as the generator is illuminated, a voltage is present at the PV generator output terminals. Installers must be aware of this.

Due to the current source characteristic of the array, overcurrent is not a problem. Under normal conditions a module or a generator can be shorted without endangering the cables. On the other hand certain fault conditions like short circuits cannot be cleared by a fuse. Thus, once a short circuit develops, it can last as long as the sun shines and might cause severe damage to the installation and even start a fire. This fault cannot be handled properly, so the installation of the generator has to be done in a way that virtually eliminates the possibility of such an incident. Using 'ground-fault-proof and short-circuit-proof' installation provides an acceptable way to cope with this hazard.

If a high system voltage is chosen, it poses the same hazard of electric shock as any conventional installation at the same voltage level. National electrical codes address this issue and should be consulted.

Partial shading of modules may cause 'hot spot' effects and can damage PV modules. To avoid this, 'bypass' diodes should be used according to the module manufacturer's specification.

Overvoltage/lightning hazard depends strongly on the location and the size of the generator. Modern PV modules are generally very rugged, but electrical equipment like inverters are usually more sensitive. If varistors for overvoltage protection (OVP) are used, attention must be paid to the possibility that ageing may lead to increased leakage currents through the device. This may eventually lead to overheating and cause a fire. Therefore, these devices should be monitored for increased leakage currents, either by an internal temperature dependent switch or by external insulation monitoring.

## 6.6 Grounding

There are two distinctly different grounding functions:

- Generally metallic enclosures of electrical equipment must be grounded to prevent a hazardous touch voltage, if an internal insulation fault develops. Similarly metallic enclosures of equipment in PV-systems should be bonded to ground. A PV generator's metallic support structure should also be grounded: no touch voltage can develop and in case of a direct lightning strike the grounded support structure may provide a convenient path for the lightning current. (If the PV generator is protected by an external lightning protection system, consideration should be given to leaving the support construction ungrounded to prevent coupling effects via ground.)
  Lightning protection devices should be directly connected to ground on a path as short as possible.

- Grounding of active parts of the PV generator is a different issue. It is possible to ground either leg of the PV generator or use a grounded center tap. This measure assures a well defined potential at the PV-generator. However, not to ground the PV generator seems advantageous in terms of reliability and personal safety. A first ground fault in a floating PV generator would not cause a hazard and the system could by left in operation. At a grounded PV generator the first isolation fault would constitute a shock hazard and should be immediately cleared.

The issue of grounding is dealt with in most countries' electricity codes. These codes should be carefully consulted before defining the grounding system for a PV generator to make sure that the local code is obeyed.

## 6.7 Accessibility and protection against electrical shock

A person touching a cell of a broken module should be protected against electrical shock (see Figure 6.4). The IEC TC 82 working group WG 3 has drafted safety rules for PV systems: 'Safety Regulations for Residential, Grid Connected PV-Power Generating Systems'. The draft is based on the standard IEC 364. It offers several 'protective measures' against electrical shock. These are:

**Safety extra low voltage, protection class III**
If the open circuit voltage of the solar array is lower than 120 V (25 °C, 1000 W/m$^2$) and the inverter has an isolation transformer, no special safety actions are necessary. No fence is needed even if the modules can be touched (e.g. at facades and motorway sound barriers).

**Protective insulation, protection class II**
If the solar modules, the array and the cabling are designed under the rules of protection class II (protective insulation) no fences and isolation transformers have to be used even when the open circuit voltage is higher than 120 V. Up to now only a few solar modules with protection class II rating are available.

**Mounting out of reach**
If the solar panels are not specified according to protection class II and the open circuit volt-

age is higher than 120 V, the modules must be installed in such a way that they cannot be reached by persons. In Germany the roof of a house is commonly accepted as such a place. Most larger PV plants are installed inside a fence because high array voltages are used.

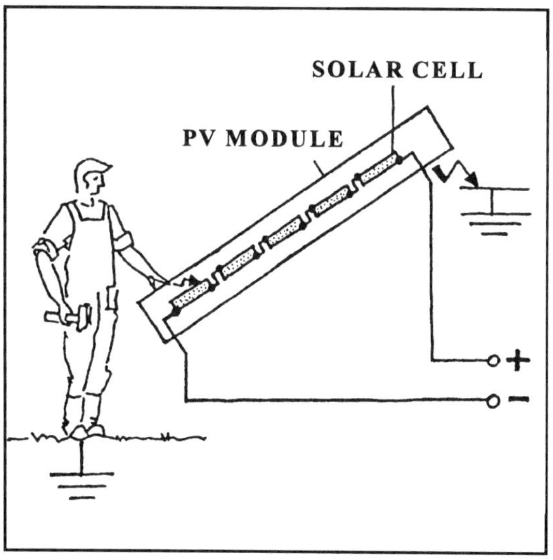


*Figure 6.4 Electrical Safety (IEC TC 82).*

**Special protection methods**

There are some situations where higher voltages ( $V_{OC} > 120$ V ) are needed, but fencing is not practicable and an enhanced protection level is desired.

- Alpine PV-projects with varying snow situations (2 - 5 m of snow);
- PV installed on motorway soundbarriers;
- PV on the facade of a building.

In this case the PV-array will be operated floating (not grounded). The inverter has to be equipped with an isolation transformer and an automatic ground fault detection device. As long as there is no ground fault any pole of the DC system can be touched without danger. If a ground fault is detected the solar generator will be shorted and the active poles ($+V_{DC}$ and $-V_{DC}$) will be connected to ground until the failure is found. At utility scale installations the ground fault message will be transmitted automatically to the headquarters by a telephone link and immediate action can be organized.

To reduce voltage stress on the modules the solar generator of large PV systems can be center tap grounded.

## Chapter 7

# Energy Storage

### 7.1 Background

A basic characteristic of sunshine is that it is variable on both daily and seasonable bases. This causes electricity production of photovoltaic modules to vary correspondingly. The mismatch between the electrical load and the electricity production must be balanced by using some kind of energy storage device.

In grid-connected PV buildings an energy storage device is usually not needed, but in off-grid houses the storage element plays an important role. The main characteristics of energy storage systems for PV building applications are cost, cycle life, availability, ease of operation and maintenance. The importance of volumetric and gravimetric energy densities will vary depending on the application. There are several energy storage possibilities from which only a few are suitable for PV-building applications. In Table 7.1 an overview of some energy storage possibilities is shown.

**Present options**

Lead-acid (Pb-acid) batteries and nickel/cadmium (Ni/Cd) batteries are the present options for PV building applications. These two types of batteries are well known and available to the consumer. There are limitations with respect to energy density, cycle life, temperature of operation and toxicity (both lead and cadmium are toxic) associated with both of these systems.

**Medium term options**

Alternative energy storage options are being developed through intense research and development. The sodium/sulphur (Na/S) system has a very high specific energy density. However, for consumer applications like PV in buildings it operates at too high temperatures. Zinc/bromine ($Zn/Br_2$) flow batteries, which have a fairly high specific energy density also, operate at ambient temperatures and necessitate flow-loop infrastructures. The installed cost of these batteries has been projected to be lower than that of improved lead-acid and nickel/cadmium batteries. Advanced lead-acid and nickel/cadmium batteries are also medium term options. Other medium term options are nickel-based batteries such as nickel/hydrogen, nickel/metal hydride, nickel/iron and nickel/zinc batteries.

Sodium/sulphur and zinc/bromine batteries are closer to commercialization than other medium term options. However, because of very good cycling performance, the nickel and nickel/metal hydride batteries have been identified as possible candidates for PV systems. The nickel/metal hydride battery could eventually replace nickel/hydrogen in metal. It can also replace nickel/cadmium systems, thereby eliminating toxic cadmium, unless the latter is recycled.

**Long-term options**

Iron/chromium redox flow batteries and rechargeable zinc/manganese dioxide batteries are long term options for PV systems.

| Battery type | Energy density | | Operation temperature | Self discharge at 20°C | Cycle life | Charge-discharge efficiency | Relative price per kWh initial cost |
|---|---|---|---|---|---|---|---|
| | [Wh/kg] | [Wh/l] | [°C] | [Capacity-%/month] | [cycles with 60...80% DOD] | [%] | |
| **Present options** | | | | | | | |
| Lead-acid (vented) | 20...45 | 40...100 | -20...+50 | 2...4[1] | 200...2000 | 70...80 | 1 |
| Lead-acid (sealed, gelled) | 10...30 | 80 | -20...+40 | 2 | 500 | 70...80 | 1...2 |
| Nickel/cadmium, pocket plate (Ni/Cd) | 15...45 | 40...90 | -20...+50 | 2 | >5000 | 60...75 | 3...5 |
| **Medium term options** | | | | | | | |
| Nickel/hydrogen (NiMH or $Ni/H_2$) | 40...60 | 60...90 | -5...+40 | 15...30 | 3000...6000 | 80...90 | 5...10 |
| Advanced Nickel-iron (Ni/Fe) | 22...60 | 60...150 | -10...+50 | 20...40 | 1000...2000 | 40...60 | 1...1.5 |
| Nickel-zinc (Ni/Zn) | 60...90 | 120 | -10...+60 | 10 | 250...350 | 75 | 2 |
| Sodium/sulphur (Na/S) | 100...250 | 150 | 300...400 | N.A. | 900...2000 | 75...90 | 0.5...1.5 |
| Zinc/bromine ($Zn/Br_2$) | 55...75 | 60...70 | -10...+50 | N.A. | 600...1800 | 70...75 | 0.5...1 |
| **Future options** | | | | | | | |
| Iron/chromium redox (Fe/Cr redox) | N.A. | N.A. | 0...+65 | N.A. | 20,000 | 60...75 | 1 |
| Rechargeable zinc/manganese dioxide ($Zn/MnO_2$) | 70 | 160 | -15...+65 | 2 | 200 | | 1...2 |
| Hydrogen storage (Fuel cell, electrolyser, gas storage)[2] | N.A. | N.A. | -20...+50 | 0 | N.A. | 35...50 | 41 |

[1] Lower value for non-antimony and high value for low-antimony batteries
[2] Economical only in large sizes (> 100 kWh energy storage) and for long term seasonal storage.

*Table 7.1 General specifications of energy storage options for PV systems.*

The combination of an electrolyzer, gas storage and a fuel cell makes an ideal energy storage for PV systems. During the summer day excess PV energy is used to power the electrolyzer, which produces hydrogen and oxygen from water. The hydrogen gas is stored in a pressure vessel. At night or in winter, when insufficient PV energy is available, hydrogen from the gas storage and oxygen from the atmosphere are fed to the fuel cell stack, in which gas to water conversion and electricity production takes place. Commercialization of this kind of energy storage systems based on hydrogen technologies is projected for the first years of the new century.

## 7.2 Lead-acid batteries

### 7.2.1 General

Lead-acid batteries have been in use for over 150 years and continue to dominate the automotive, power backup and traction markets. The main reason for this is their relatively low cost and their long and reliable service life. They come in many shapes, sizes, types and designs depending on the application. They can be sold as single cells or more commonly as a series combination of six cells to form a 12 V battery. Different lead-acid battery models are shown in Figure 7.1.

Batteries are often sold by their amp-hours (Ah) of nominal capacity: car batteries by their 20 hour capacity and power backup batteries by their 8-hour capacity. This indicates how much energy can be removed during a single discharge. Small PV systems, up to about 120 Ah, mainly use modified car or truck batteries. Unfortunately, certain types of car batteries are unsuited particularly for solar applications and can have a rather short service life (1-2 years). This is because they have been designed to give high currents for starting a car or truck but not for cycle duty as required in solar applications.

*Figure 7.1 Different lead-acid battery models (NAPS).*

Until recently all batteries were of the flooded design with the plates and separators totally immersed in acid. For certain remote solar applications, batteries are designed in which additional acid is added to the container to lengthen the time between addition of distilled water. In order to overcome the trouble of periodically adding distilled water, batteries are also being produced as gas recombination batteries. These are sealed lead-acid batteries also known as valve regulated batteries. Here the gas produced during charging recombines in the cell to form water.

### 7.2.2 Structure and principles of operation

All lead-acid batteries have the same general structure. The main components of a single cell are positive and negative plates, terminals, separators, sulphuric acid ($H_2SO_4$), container, terminal sealings and a safety plug. The plug is used mainly to minimize acid mist, but also to decrease the volume of escaping gases generated by gassing reactions occurring at later stages of charging. In Figure 7.2, the general structure and the main components of a single cell stationary battery are shown.

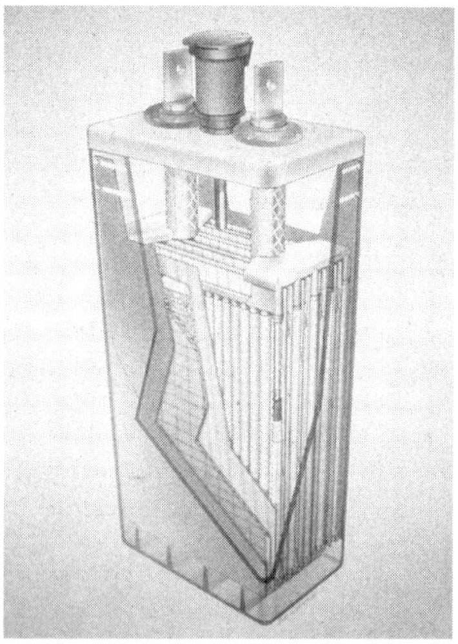

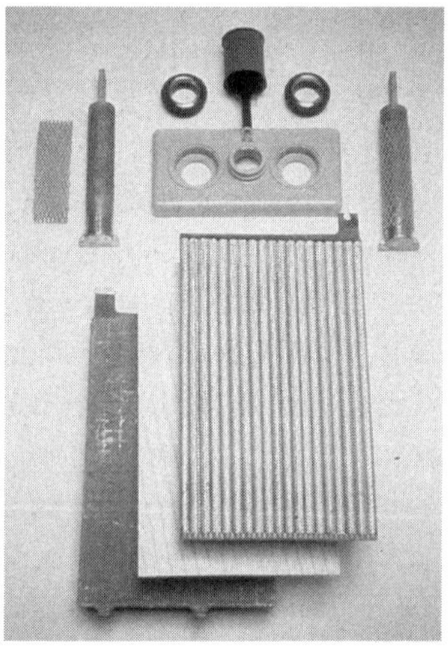

*Figure 7.2 Single cell tubular plate stationary lead acid battery (NAPS)*
*a) General structure (top)*
*b) Main components (bottom).*

The negative plate or anode is composed of a negative grid pasted with sponge lead (Pb), the positive plate or cathode is a positive grid pasted with lead dioxide ($PbO_2$). The electrolyte is sulphuric acid ($H_2SO_4$) in water solution, which also participates in the discharge reaction according to:

$$Pb + PbO_2 + 2\ H_2SO_4 \rightarrow 2\ PbSO_4 + 2\ H_2O\ .$$

The reverse reaction occurs during charging, sulphuric acid is formed as lead sulphate and is converted into lead and lead dioxide.

The acid concentration in lead-acid batteries varies depending on its use. For car batteries it is approximately 38% (specific gravity 1.28 kg/dm$^3$), while in standby batteries it can be as low as 32% corresponding to a specific gravity of 1.22 kg/dm$^3$. The nominal cell voltage is 2 V, but the actual open circuit voltage of a fully charged cell is in the range of 2.1 V...2.4 V depending on acid density and temperature. During discharge, the operating cell voltages decrease from above average to the cut-off between 1.75 V and 1.9 V. This cut-off voltage is very important for the battery lifetime as it defines the depth-of-discharge (DOD). The DOD is both rate and temperature dependent (see section 7.5) At very low rates, the battery capacity can actually be higher than the 20 hour or eight hour nominal capacity thus giving a DOD of greater than 100%. Therefore, at low rates the cut-off voltage can and should be quite high. The capacity increase at low discharge rates has two causes. The first is that the fine pores in the plates do not block as $PbSO_4$ is formed and acid can penetrate deep into the plates. The second reason is that the voltage drop across the battery itself, which is the product of current and the battery's internal resistance, is rate dependent. At low rates, the voltage drop is low and the capacity can be high. During charging the applied voltage must be in the range of 2.3 V...2.5 V per cell in order to charge the battery in a reasonable time. During the later stages of charging, gassing occurs producing hydrogen and oxygen from water. This has the beneficial effect of stirring up the electrolyte and avoiding acid stratification. However, gassing increases the water consumption and also needs routine maintenance.

Acid stratification is caused by the continuous deep discharging and charging of batteries without stirring of the electrolyte between cycles. It is especially problematic in tall cells (> 60 cm) where the acid density can range from > 1.4 kg/dm$^3$ at the bottom to < 1.2 kg/dm$^3$ or less on top. This causes non-uniform discharging of the plates, reduces capacity and shortens operating life.

The grids of lead-acid batteries are normally composed of lead plus a variety of metals in concentrations ranging from 0.1% up to 5-8% of the grid weight. Battery grids can be made of pure lead, but due to the softness of the metal, a special manufacturing technology must be applied. Antimony is used in positive grids from 0.5% to 8% to strengthen the grids and to improve its cycling characteristics. But antimony will also increase the gassing reactions and self-discharge. Small amounts of calcium (0.1-0.7% of the grid weight) can also be used to improve grid hardness but it does not improve the cycling performance of the battery as antimony does. However, a battery with calcium has the advantage of lower self-discharge and less gassing than an antimony battery. Other metals such as tin, arsenic and silver may also be added to improve castability, metallurgical and mechanical properties.

### 7.2.3 Lead-acid battery classification

Lead-acid batteries can be classified according to numerous methods such as: sealed or non-sealed; by application: car, power backup, traction; by type of positive plate: flat or tubular. For PV applications a useful classification is:

- antimony (deep cycle)
- non-antimony (shallow cycle).

Antimony batteries cannot be hermetically sealed because antimony promotes the gassing reactions that occur at the end of charging. But antimony can be used with other types of batteries. Tubular plate batteries are often used for cycling purposes with antimony concentrations up to 8% (traction batteries) but are more expensive than flat plate batteries and are usually not available for consumer use. However, PV applications typically have shallow cycles and therefore grid alloys with 1...3% antimony are sufficient. With higher antimony levels gassing and maintenance needs increase too much for most PV applications.

The non-antimony batteries are typically made with calcium alloys and the main advantage of these is reduced maintenance and low self-discharge causing long shelf life. Flooded type batteries must always have a small opening to let the gases escape. Those gases contain hydrogen and oxygen and can be explosive and harmful. If the battery is made as sealed type, the gases formed during charging will recombine in the battery to form water. In order to achieve rapid recombination of the gases, tiny gas passages must be formed between the plates. This is done either by forming an electrolyte of silica gel or by using glass fibre separators. The disadvantage of these batteries, besides the fact that they do not cycle as well as those containing antimony, is that it is very important to limit the charging voltage near the end of charge to under 2.35 - 2.4 V per cell. At higher voltages the gassing reaction will be faster than the recombination reaction and the battery will dry out due to the escaping gases. This voltage limit will lengthen the charging time.

### 7.2.4 Factors affecting lead-acid battery life

Service life of a PV battery can be determined by cycle life (if the regular cycling is quite deep) or by positive grid corrosion (if the regular cycling is quite shallow). Normal car starter batteries have thin plates, and both cycle life and resistance to corrosion are low compared with other types of lead-acid batteries. In many cases, especially where the service life is controlled by positive grid corrosion, flat plate batteries with thick lead grids can have an expected service life approaching that of tubular plate batteries.

The useful service life of a battery is normally considered completed when the battery can no longer do the job. This is often when the capacity has decreased to less than a certain percentage (e.g. 80%) of the nominal capacity. The most common failure mode of a battery is the gradual loss of capacity caused by a combination of active material degradation due to cycling and grid corrosion. But other failure modes are possible such as cell shorts, cracked cases or broken plates, which lead to a sudden loss of capacity. The most common factors affecting battery life are listed below:

- deep (>50%) daily discharging, which causes positive plate shedding (sludge formation at the positive plate),
- high temperatures that speed up corrosion,
- prolonged overcharging which increases the rate of corrosion,
- prolonged undercharging leading to sulphation (white spots) and acid stratification which reduces plate capacity,
- antimony levels of 1...3% which increase cycle life.

Other factors such as a low electrolyte liquid level will also severely shorten the battery lifetime.

Sulphation occurs after a battery has been deeply discharged which in PV applications could occur during prolonged cloudy periods. Here lead-sulphate crystals which are always formed during battery discharge begin to grow and convert slowly into a form that is difficult to recover. Permanent loss of capacity can occur. Additionally, after a battery has been continuously deeply discharged, acid stratification takes place. This can be avoided by overcharging the battery thus forming gases which stir up the acid. In larger systems, air lift pumps can be used to ensure uniform acid density.

## 7.3 Nickel/cadmium batteries

### 7.3.1 Principles of operation and characteristics

In nickel/cadmium batteries, hydrated nickel oxide (NiO) is the cathode and cadmium (Cd) is the anode during discharge. A potassium hydroxide water solution is used as the electrolyte. The cell discharge reaction is shown below:

$$2NiO\text{-}OH + Cd + 2H_2O \rightarrow 2Ni(OH)_2 + Cd(OH)_2 \,.$$

The reverse reaction occurs during charging. However, the reactions are not nearly as simple as indicated particularly at the positive electrode. This is illustrated by the shape of the charging curve presented in Figure 7.3.

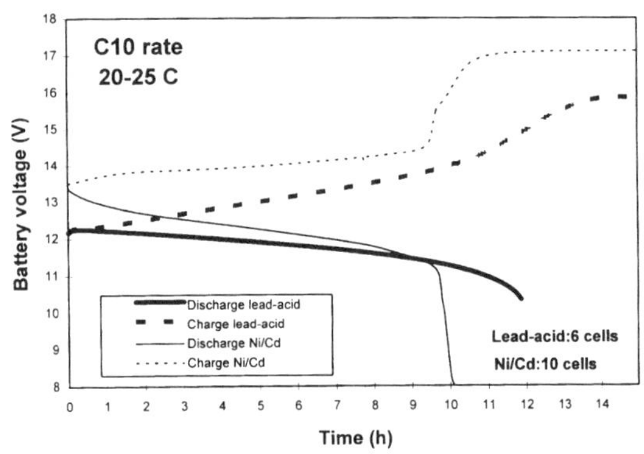


*Figure 7.3 Nickel/cadmium and lead-acid battery voltage during charging and discharging at 25° C and with 10 h rate.*

Nickel/cadmium batteries are mechanically rugged and have a long cycle life. They have better low temperature characteristics down to -20° C. Problems of electrolyte stratification or sulphation as in lead-acid batteries do not occur in nickel/cadmium batteries. The nominal cost per Ah is 3-5 times higher than for lead-acid batteries. The battery can be cycled deeper. High cycle life and the capability to be operated at low temperatures partly compensate for the higher investment. The rapid voltage increase during the final stage of charge indicates that the energy efficiency is low if the battery is fully charged. In practice, because electrolyte stirring is not needed as in lead-acid batteries, Ni/Cd batteries do not always have to be fully charged. Typically the end of discharge voltage is 1.0 V/cell.

### 7.3.2 Ni/Cd types available

Nickel/cadmium batteries are available in sealed or vented types. Vented types are made as sintered plate or pocket plate construction. A sintered plate is fabricated by impregnating the active material into a nickel support plate. With this design, a lower internal resistance and sensitivity to variable temperature operation can be achieved than with a pocket plate construction. Here the active material is contained in perforated pockets and features a more rugged plate structure than the sintered type. Batteries with pocket plate design have extended cycle life and the ability to withstand extended periods at partial state of charge without sustaining damage. Sintered plate cells have a tendency to suffer from a 'memory effect' phenomenon. This effect is caused by repeated incomplete discharge, which eventually results in capacity reduction, since the 'remembered' capacity appears to be smaller than the actual one. This temporary effect can sometimes be eliminated by subjecting the cell to occasional single deep discharge/charge cycle.

Sealed nickel/cadmium batteries were first developed in the 1950s from vented sintered plate nickel/cadmium batteries. The original sealed batteries used the same active material and similar intercell components as those used in vented batteries. Since this time, materials and production techniques have improved, enabling the sealed battery performance to improve, both in terms of charge retention and discharge capacity.

Vented pocket plate Ni/Cd batteries are produced in capacities from 10 to 1200 Ah single cells and monoblocks, while commercially available sealed sintered plate cells range mainly from 0.1 to 23 Ah in sizes of typical primary batteries. Vented sintered plate cells are made up to sizes of 1000 Ah of single cells and monoblocks.

### 7.3.3 Factors affecting nickel/cadmium battery life

In sintered nickel/cadmium batteries, deep-discharging is actually beneficial because the 'memory effect' that results from shallow discharge can be avoided. Improvements in the metallurgical structure due to plate fabrication and design are important issues in controlling the 'memory effect' and cycle life of the nickel/cadmium batteries. Electrolyte agitation is unimportant for nickel/cadmium batteries, but fouling through carbonation (caused by $CO_2$ in the air, e.g. batteries left near the exhaust of a diesel generator) of the alkaline electrolyte decreases the battery life.

## 7.4 Safety aspects

Environmental aspects of batteries should be considered from the point of view of the life-cycle of the battery. Environmental safety, i.e. pollution prevention is an important issue. For all batteries, recovery of battery degradation products after their useful life should be controlled.

Battery safety during operation is also important. Gassing during charging produces hydrogen and oxygen gas, which under certain conditions can lead to explosion or fire. In order to avoid this, vented batteries should be placed in a well-ventilated area and away from other electrical components that could produce a spark. The required ventilation depends on the size of the battery and its charging rate. Sealed batteries with low discharge/charge rates can often be stored inside without ventilation, since most buildings are not airtight. However, as a general principle, it is advisable to provide ventilation even when sealed lead-acid batteries are used. Also, batteries should be placed in cool and dry surroundings where possible.

Vented lead-acid and nickel/cadmium batteries are subject to explosion hazard if the venting valves become plugged. Periodic maintenance can prevent this hazard.

Cell imbalance is also a safety issue and the imbalance occurs when one or more cells in a series string are weaker than the rest, which can lead to cell shorting. It is important to monitor annually the individual cell performance to identify weak or shorted cells. Cells, which are clearly weaker than the average should be removed from the series string since they lead to overcharging of functioning cells. This condition will cause unnecessary gassing. Modern methods to avoid all imbalances are treated in Chapter 8 (e.g. CHarge EQualizing).

Batteries can also be an electrical shock hazard. Insulated tools should always be used. DC isolation switches should be used when working on other parts of the system. Service technicians must be trained about the hazards associated with large battery storage units. The electrical power available from a large storage bank is very high, a short circuit current of 1000 A to 4000 A is readily obtainable from a single cell.

The electrolytes used in most PV system batteries are corrosive. Sulphuric acid in lead-acid batteries, and potassium hydroxide in Ni/Cd batteries can cause burns when in contact with skin. A small amount of electrolyte in the eye can result in blindness if not immediately flushed with water. The use of safety glasses or safety face shields should be mandatory while performing maintenance on these batteries.

## 7.5 Interpretation of the manufacturer's data sheets

Battery manufacturers provide general information such as cycle life, float life, battery capacity, battery voltage and operational temperature range along with cost upon request. Field experience indicates that this information cannot necessarily be applied directly to PV applications. Typical parameters like cycle life and capacity are measured under controlled laboratory conditions. However, in PV applications the operating conditions are quite variable. There is no standard cycle and both high temperatures and poor charging can reduce cycle life by 10% to 50%. The battery capacity (in Ah) also depends upon discharge rate, temperature and cut-off voltage. Thus, performance comparison of batteries of different manufacturers on the basis of their data sheets is not straightforward.

An important battery parameter is its **nominal capacity (Ah)**. However, the **usable capacity** of a battery is more important than the nominal capacity and it depends upon how the battery is being charged and discharged. The usable capacity decreases with an increase in discharge rate and decrease in temperature. Usually, if more storage capacity is required, bigger cells are used, and if higher voltage is required, they are connected in series. Batteries are seldom connected in parallel.

The **discharge rate** of a battery is expressed in amperes or in discharge time.

$$Rate(A) = \frac{Capacity(Ah)}{Discharge\ time(h)}$$

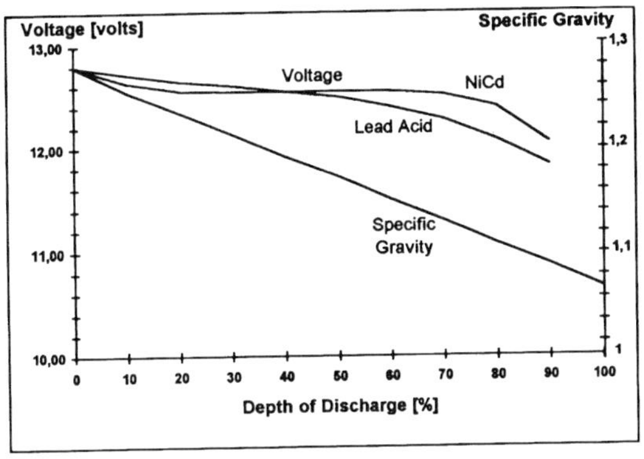


*Figure 7.4 Effect of depth of discharge (DOD) on Ni/Cd and lead-acid battery voltage and specific gravity of lead-acid batteries at a certain temperature during discharge./1/*

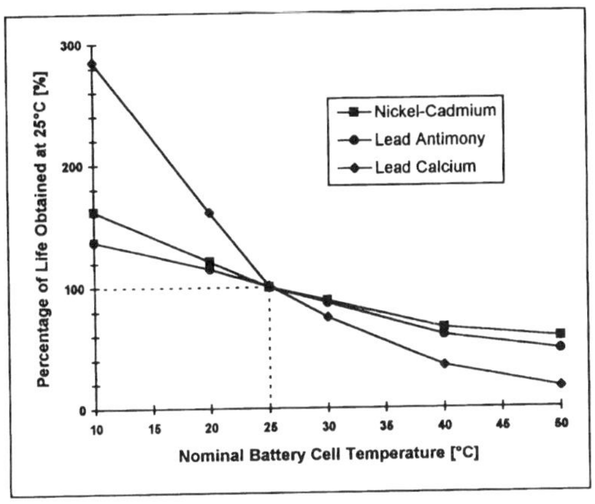


*Figure 7.5 Effect of temperature on lead-acid and nickel/cadmium battery life./1/*

For example, if a battery is discharging at a rate that would take 50 hours to fully discharge to the 'cut-off' voltage, then the battery is considered to discharge at $I_{50}$ current or '50 h' rate ($C_{50}$). For typical PV applications a $C_{100}$ rate is desirable but as low as $C_{500}$ is possible in some cases.

**The battery voltage** depends on the type of battery, state of charge, rate of discharge or charge, temperature and acid density. It is higher while charging than during discharging. Figure 7.4 shows how the specific gravity and voltage of a lead-acid battery and of a nickel/cadmium battery voltage vary with depth of discharge or state of charge during discharge.

Figures 7.5 and 7.6 show the effect of temperature and discharge on delivered capacities and lifetime of lead batteries. It can be seen that with an increase in temperature battery life decreases more with sealed than with flooded lead-acid. Higher discharge rates and low temperatures decrease the delivered capacity. The temperature and discharge rate do not affect the Ni/Cd battery capacity as much as they do lead-acid batteries.

Battery manufacturers' data sheets provide a general overview or a range of characteristics under an envelope of operating conditions. Hence, testing of the candidate batteries under predicted or simulated PV service conditions is quite important. If this is not possible, the data provided by the manufacturer should be used and extrapolated for the PV conditions.

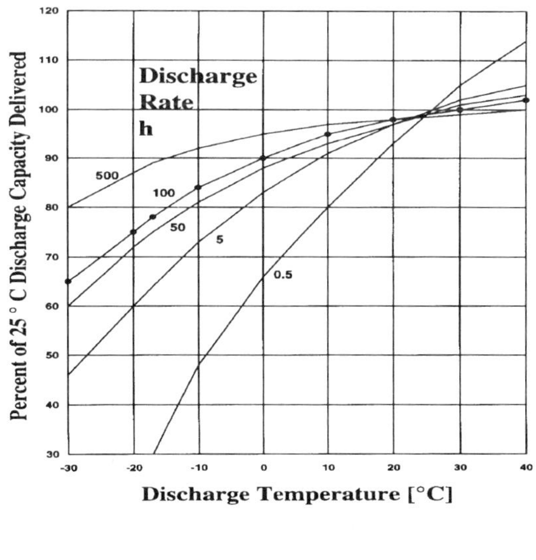


*Figure 7.6 Effect of discharge rate and temperature on capacity of lead-acid batteries./1/*

/1/ Canadian Phototvoltaic Industries Association, *Photovoltaic Systems Design Manual*, CANMET-Energy, Mines & Resources Canada, Canada, 1991. Reproduced by courtesy of CANMET/NRCan.

## Chapter 8

# DC Power Conditioning

### 8.1 Introduction

The solar generator, the heart of a PV system, fundamentally produces DC currents and DC voltages. This DC power can either

- be used without intermediate energy storage by directly coupled consumers like electric fans or circulation pumps in solar thermal systems,
- be stored in a battery and supply stand-alone systems like small appliances, lamps or even remote houses or
- be fed into the public grid by DC to AC converters.

In almost all cases, special power conditioning units are required for:

- the optimal operation of the solar generator and
- the optimal and safe operation of the connected electrical equipment.

Depending on the application, the following power conditioning units (PCUs) may be needed in a PV system:

- DC to AC converters (inverters),
- matching DC/DC converters (MCs),
- charge controllers.

Considering PV applications in buildings, the inverter is one of the key components. For this reason, inverters for grid-connected as well as for stand-alone systems will be discussed in detail in Chapter 9.

In this chapter, the question of how to connect PV generators to different loads will be answered.

### 8.2 Matching of solar generator and different loads

#### 8.2.1 General

The power produced by a solar generator in a given operating point can be calculated by multiplying the corresponding current and voltage. Doing this from short circuit to open circuit conditions at constant insolation leads to the power curve shown in Figure 8.1.

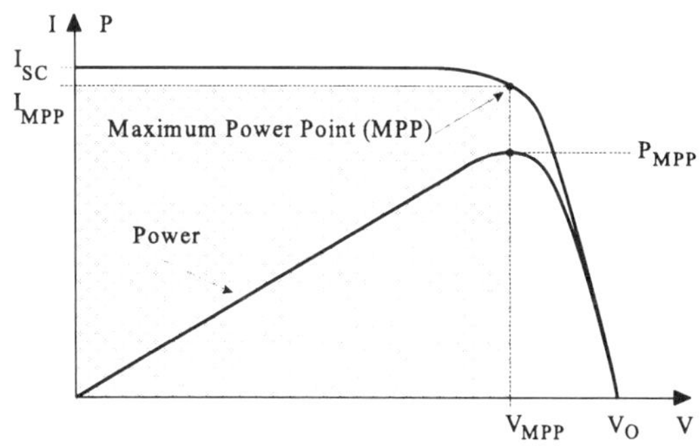


*Figure 8.1 IV curve and power curve for constant insolation.*

The output power is zero when the terminals of the generator are short-circuited or left open, and between these two extreme operating points there is one combination of current and voltage on the IV curve where their product (the resulting rectangular area) is maximal. This special operating point is called the Maximum Power Point or MPP. To gain as much energy as possible, the solar generator should theoretically be operated close to the MPP under all conditions of insolation, temperature and changing loads. However, experience shows that the energy gained by tracking the MPP is often overestimated, especially when the losses in the PCU due to self-consumption and incorrect operation are taken into account. Although many PCUs with MPP tracking capability are available on the market, the value of such an additional component should be evaluated thoroughly.

The following three examples give some general guidelines.

### 8.2.2 Coupling of PV generator and batteries

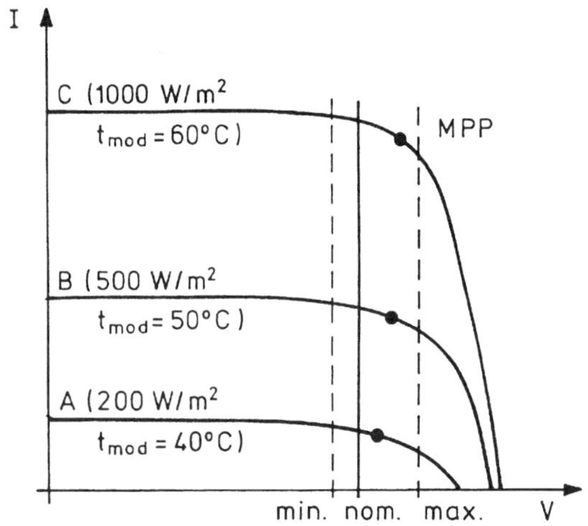


*Figure 8.2 Typical PV generator IV curve for three different insolation values.*

Figure 8.2 shows a typical PV generator IV curve for three different insolation values. In contrast to the IV curves normally given in data sheets, here the influence of increasing module temperature with increasing insolation for a typical way of module mounting has been considered. Furthermore, the IV curve of an ideal battery (a vertical line) is shown, which varies within a given voltage range depending on the state of charge (e.g. 11 V to 14.4 V for a 12 V lead-acid battery).

From Figure 8.2 it can clearly be seen that during charging, the PV generator is operated close to the MPP without any additional control if the nominal voltage of the PV generator is chosen appropriately (in practice 33....36 crystalline silicon cells for a 12-V lead acid battery). This statement applies also for direct coupling of PV generators and electrolysers.

### 8.2.3 Coupling of PV generator and grid-connected inverter

The operating voltage of a PV generator connected to the input of a grid-connected inverter is determined by the control algorithm of the inverter. Therefore, the input characteristic of a grid-connected inverter is a vertical line, as above. The terminal voltage of a battery is determined by its state of charge and cannot be influenced by the operator or by a control algorithm. In contrast, the input voltage of a grid-connected inverter can easily and instantaneously be shifted by an MPP-Tracker implemented either in the hardware or the control software. In general, there is no need for additional power electronics that cause extra energy losses. From this point of view, MPP tracking can be recommended for grid connected inverters.

### 8.2.4 Coupling of PV generator and DC motors or ohmic loads

The IV curve of fans or pumps driven by DC motors can be approximated by sloped lines as shown in Figure 8.3 for three different ohmic loads.

Load curve 2 is optimally matched to the IV curve B at medium insolation, but at lower and higher insolation (curves A and C, respectively) there are considerable energy losses due to mismatch (the operating points are far away from the MPP). In all three cases, there is a clear mismatch at most insolation values - in many cases the best compromise will be system dimensioning according to curve 2. To overcome the mismatching problem fundamentally, a matching DC/DC converter (MC) is switched between the PV generator and the load as shown in Figure 8.4.

It is the task of this MC to keep the operating point of the PV generator close to the MPP under all operating conditions. In contrast to normally used DC/DC converters where the output voltage is controlled and stabilized, in this application the input voltage (PV generator voltage) is kept at a constant value. This voltage can either be manually adjusted by the user or automatically track the MPP by an appropriate control algorithm. The output voltage will vary depending on the input power and load characteristics.

Using an MC can cause a considerable increase in power, especially at low insolation levels, which leads to a much earlier start of pumps and fans etc. Therefore, MCs are often already built into pumps or available as extra components on the market under trade marks such as 'Maximizer', 'LCB' (**L**inear **C**urrent **B**ooster) or 'APW' (**A**n**p**aß**w**andler). Some of these units provide additional features such as an adjustable output voltage limiter ('dimmer') or remote control inputs for submersible pumps.

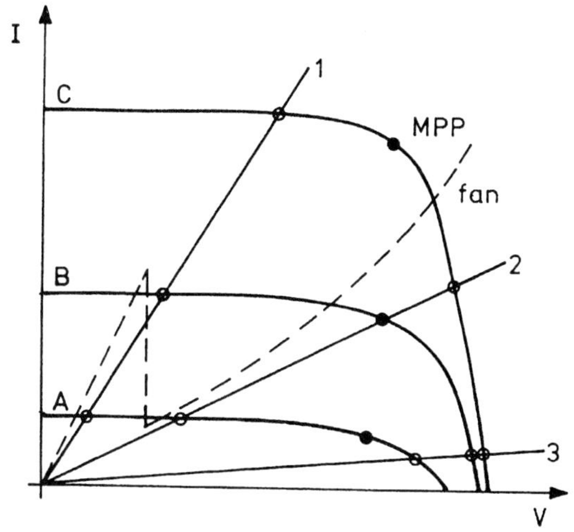


*Figure 8.3 IV curves of PV generator, electric fan (dotted line) and three different ohmic loads.*

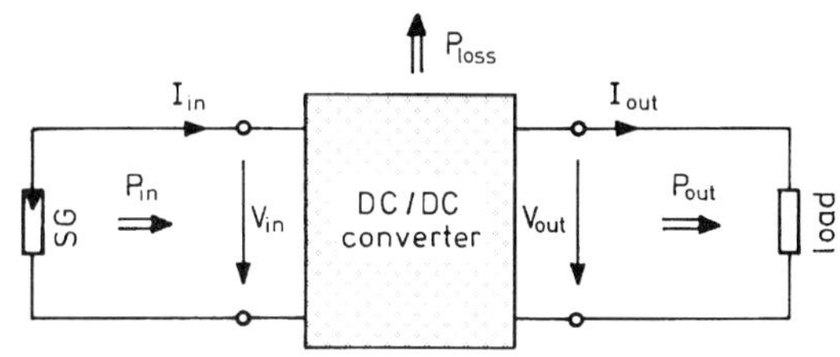


*Figure 8.4 Matching DC/DC converter.*

## 8.3 Maximum Power Point Tracking (MPPT) algorithms

Developing MPPT algorithms is a challenge for every engineer, so a large variety of different types has been invented. The most used algorithm is able to find the MPP of the IV curve under normal operation. To find the MPP, the operating voltage of the PV generator is periodically (e.g. every few seconds) changed by a small amount. If the output power of the PCU increases due to this change, the operating voltage will be changed in the same direction at the next step. Otherwise, the search direction will be reversed. The operating point thus swings around the actual MPP. This very simple algorithm can be improved, e.g in order to suppress steps in the wrong direction due to rapidly changing insolation.

## 8.4 Charge controllers

### 8.4.1 General

Most of today's PV applications require energy storage due to time shifts between the energy supply and demand. In a typical power supply for remote houses, this energy storage represents approximately 15-20% of the initial cost. Taking into account that during the system lifetime the storage batteries have to be replaced several times, the cost share for batteries can exceed 50% of the total costs over the system lifetime. Experience shows that in most PV applications the battery lifetime is much shorter than expected - sometimes only 2-4 years instead of 5-10 years as often stated by suppliers. The goal of worldwide intensive research work is to extend the battery lifetime and to develop peripheral battery hardware that is suited to the battery demands.

The main task of a charge controller is to operate the battery within limits defined by the battery manufacturer regarding overcharging and deep discharging. Furthermore, a charge controller can take over automatic and regular 'maintenance duties' like equalization charging or overcharging to prevent acid stratification. Advanced charge controllers incorporate a monitoring system that informs the user about the state of charge of the battery and the battery history (e.g. number of deep discharge periods, Ah balance). In larger hybrid systems, the charge controller acts as an Energy Management System (EMS) that automatically starts back-up generators (e.g. diesel gensets) as soon as the battery's state of charge drops below defined limits.

The fundamental principles of state-of-the-art charge controllers as well as an outlook on future developments is presented in the following section.

### 8.4.2 Overcharging protection

As already explained in Chapter 7, batteries must be protected from extensive and prolonged overcharging by reducing the charging current.

Three fundamental principles are commonly used:

**Series controller**

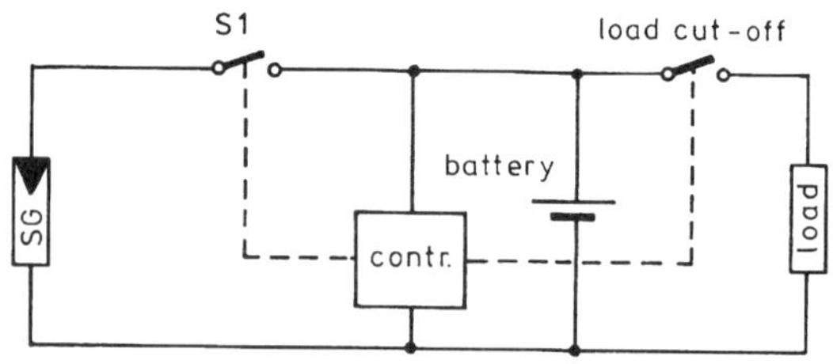


*Figure 8.5 Series Charge Controller.*

To stop the charging process, a switch (relay or semiconductor switch) is connected in series with the PV generator as shown in Figure 8.5. As soon as the battery voltage has reached the end-of-charge voltage, the switch is opened by means of a controller.

One advantage of the series controller is that in addition to PV generators, other energy sources such as wind turbines can also be connected to the input. A disadvantage can be (depending on the circuit design) that the charging process cannot be started if the battery is fully depleted (0 V) because there is no energy to operate the series switch.

**Parallel or shunt controller**
A shunt controller as shown in Figure 8.6 makes use of the fact that a PV generator can be operated in the short circuit mode for any time without damage.

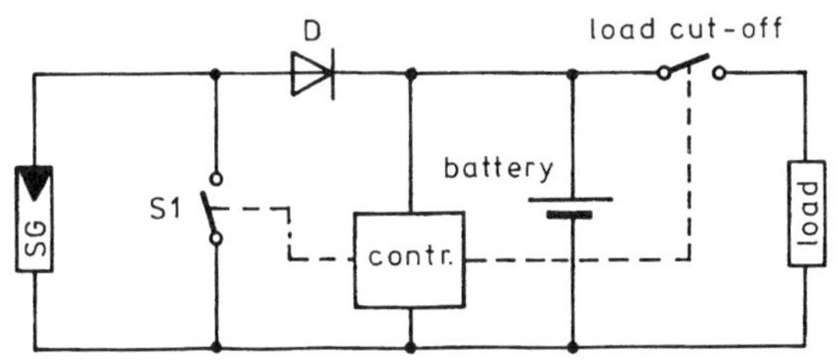


*Figure 8.6 Shunt controller.*

While charging, the current flows through the blocking diode D into the battery. When the end-of-charge voltage is reached, the PV generator is short-circuited by the switch $S_1$. The blocking diode now prevents reverse current flowing from the battery into the switch. Furthermore, it suppresses discharging currents into the PV generator during the night.

In contrast to the series controller, this kind of charge controller will also reliably start charging with a fully discharged battery, because the switch has to be energized only when the battery is fully charged. Most of the charge controllers available on the market are based on the shunt principle.

When the end-of-charge voltage is reached for the first time, the battery is not yet fully charged. The missing 5 - 10% of charge can be added to the battery by keeping it at the end-of-charge voltage level for a prolonged period while the charging current slowly decreases. How can such a charging regime be implemented by the series and shunt controllers described above, which can only switch the full PV generator current on and off? The technique used to achieve this behaviour is Pulse Width Modulation (PWM) as shown in Figure 8.7.

As soon as the battery voltage reaches the end of charge voltage, the charging current is dropped to zero by either opening the series switch or closing the parallel switch. As a result, the battery terminal voltage decreases. The charging current is enabled again when the battery voltage drops below a threshold that is approximately 50 mV/cell lower than the end of charge voltage. This sequence repeats periodically and while the charging pulses become shorter and shorter with increasing state of charge. The average charging current decreases while the terminal voltage is more or less constant.

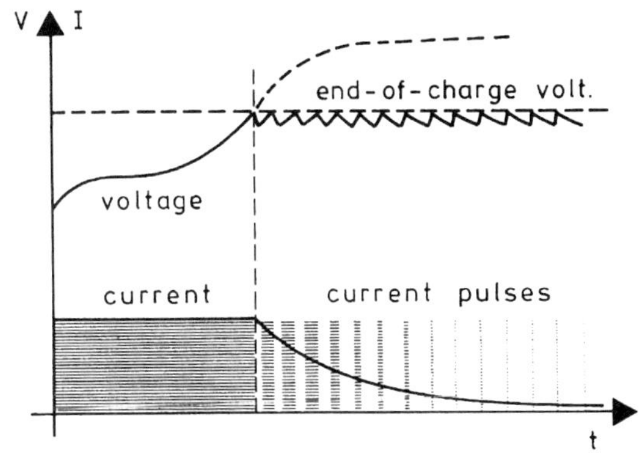


*Figure 8.7 Battery voltage and current during charging.*

**MPPT - Charge controllers**

Both the battery voltage and the PV generator voltage vary during operation due to the changing state of charge and boundary conditions such as temperature or insolation. In principle, this leads to a certain mismatch between the battery voltage and the optimum PV generator voltage and therefore causes energy losses. These losses usually are overestimated. If the components are chosen appropriately as shown in Figure 8.2, the energy losses are in the range of a few percent when using direct coupling via series or shunt controllers compared to ideal matching by DC/DC converters and MPPT! Experience shows that charge controllers based on DC/DC converters are not relevant when PV systems' energy output is to be maximized!

Nevertheless, there are two advantages in using such charge controllers:

- there is a greater flexibility in selecting modules and batteries and
- in case of very long wires from PV generator to battery, much higher generator operating voltage can be chosen than the battery voltage, resulting in lower currents and wiring losses.

The necessity of a matching DC/DC converter has to be checked carefully in any case.

**Charging strategies**

Simple charge controllers are equipped with only one threshold for the end-of-charge voltage. This threshold should be adjusted to 2.3 V/cell for lead-acid batteries at a cell temperature of 20 °C. If the cell temperature differs more than 5 K from this reference temperature, the end-of-charge voltage has to be corrected by -4 to -6 mV/K according to the battery manufacturer's recommendations. More highly sophisticated chargers provide several voltage thresholds combined with timers. They therefore allow e.g. regular overcharging (gassing) of the battery to prevent stratification of the electrolyte. These intervals are based on experience, e.g. gassing every four weeks and after each deep discharge cycle. The total gassing time should be restricted to 10 hours per month. During charging, the end-of-charge voltage should be limited to 2.5 V/cell. After that the float voltage should be 2.25 V/cell.

Attention: Never overcharge sealed (maintenance-free) batteries!

### 8.4.3 Prevention of deep discharge

To achieve maximum service lifetime of lead-acid batteries, deep discharge cycles as well as prolonged periods in partially charged conditions should be prevented. When the deep discharge threshold is approached, the load should be disconnected. The load cut-off voltage depends on the battery type and the discharge current. In PV applications it should be relatively high, e.g. 1.80 - 1.85 V/cell. To allow high, but short discharge currents, e.g. to start refrigerators, an appropriate time delay $t_d$ must be incorporated as shown in Figure 8.8.

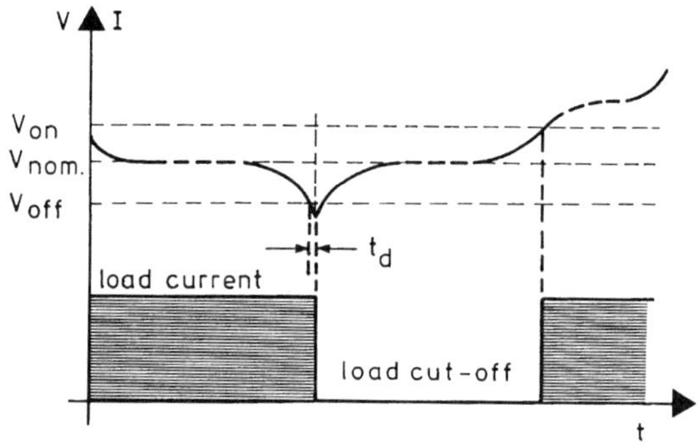


*Figure 8.8 Voltage thresholds for deep discharge protection.*

A very important feature is that the load should be connected to the battery after deep discharge only if an adequate amount of charge has been fed into the battery, i.e. the battery voltage is above 2.2 V/cell.

### 8.4.4 Future trends

Experience shows that the battery is the 'weak link' in a PV system. In the future, specially designed PV batteries should be provided by the manufacturers that withstand deep cycling and partial charging. Furthermore, the simple charge controllers used today should become more 'intelligent'. One aspect is that a battery is a varying system that changes its major parameters over its lifetime. Intelligent charge controllers recognize these variations and adapt themselves to the ageing battery. In laboratories, there are first examples of such controllers and energy management systems based on neural networks and 'fuzzy logic'.

Another attempt to extend battery lifetime is the use of so-called CHarge EQualizers (CHEQs). Conventional charge controllers assume that all cells of a high voltage battery string are in the same state-of-charge and therefore simply monitor the overall terminal voltage. In practice, however, each cell has its individual characteristics such as capacity, self discharge etc. This non-ideal behaviour of the cells may lead to large inhomogenities in the state-of-charge of cells causing deep discharge or even inverse charging of weak cells or overcharging of other cells.

The new CHarge EQualizer system prevents these inhomogenities by charge transfer between individual cells as shown in Figure 8.9.

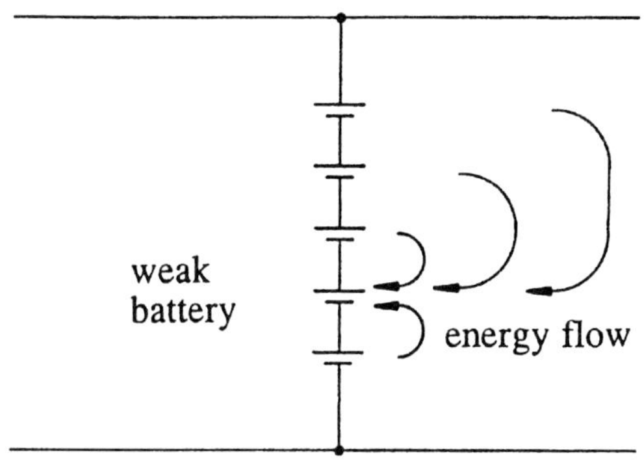


*a)* *Support of weak cells by strong cells.*

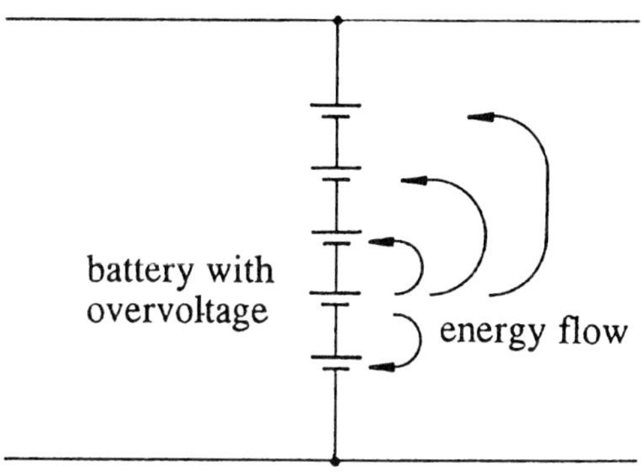


*b)* *Energy transfer from fully charged cell to remaining cells.*

*Figure 8.9* *Operating principle of CHarge EQualizers.*

## Chapter 9

# Inverters

### 9.1 General

A photovoltaic (PV) array, regardless of its size or sophistication, can generate only direct current (DC) electricity. Fortunately, there are many applications for which direct current is perfectly suitable. Charging batteries, for example, can easily be done by directly connecting them with a solar module. Inverters are required in systems which supply power to alternating current (AC) loads or feed PV electricity to the utility grid.

Inverters convert the DC output of the array and / or battery to standard AC power similar to that supplied by utilities. Inverters are solid state electronic devices. Broadly speaking, these inverters may be divided into two categories:

- stand-alone and
- utility-interactive (or line-tied).

Both types have several similarities but are different in the layout of the control circuit.

The stand-alone inverter is capable of functioning independently of the public utility grid. The correct timing of the 50/60 Hz AC output is done by an internal frequency generator.

Utility-interactive inverters have the added task of integrating smoothly with the voltage and frequency characteristics of the utility-generated power present on the distribution line.

For both types of inverters the conversion efficiency is a very important consideration (should be more than 90% for $P/P_n > 0.1$).

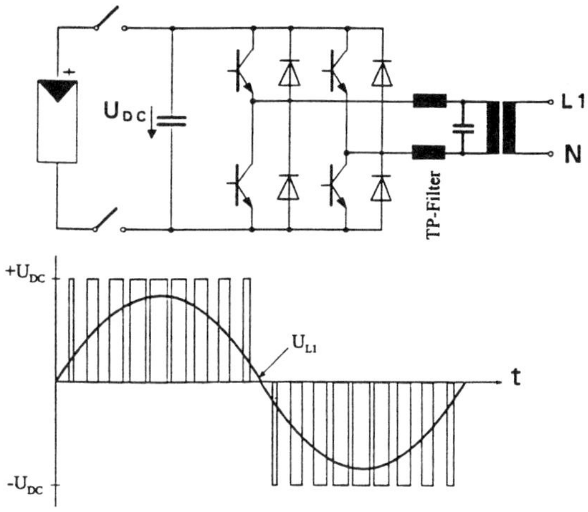


*Figure 9.1 The full bridge inverter circuit.*

### 9.2 Inverters for stand-alone applications

In many stand-alone photovoltaic installations alternating current is needed to operate common 230 V (110 V), 50 Hz (60 Hz) household appliances. Therefore, DC electricity from the battery bank has to be converted to alternating current. The conversion is done by a device called stand-alone or self-commutated inverter (see Figure 9.2). High conversion efficiency is essential for the use in autonomous systems with battery storage. Common stand-alone inverters operate at 12, 24 or 48 V DC.

The shape of the output waveform is an indication of the quality and cost of the inverter. In general it is advisable to install sine wave inverters. By using a control unit with pulse width modulation the switches in Figure 9.1 can be operated in such a way that a sine wave is shaped. The output signal can be improved further by using a low pass filter at the inverter output. For the operation of sensitive loads the harmonic content of the output voltage should be low (THD - non 50/60 Hz oscillations - lower than 3 - 5%, cost: 1.5 US $/W, 1994).

For some applications a square-wave inverter can be used. This device is based on a 50/60 Hz switched full bridge circuit and is a lot cheaper than a sine wave unit. Conversion efficiency is good but the harmonic content is much higher. Some appliances can be overheated or damaged when connected to a square-wave inverter.

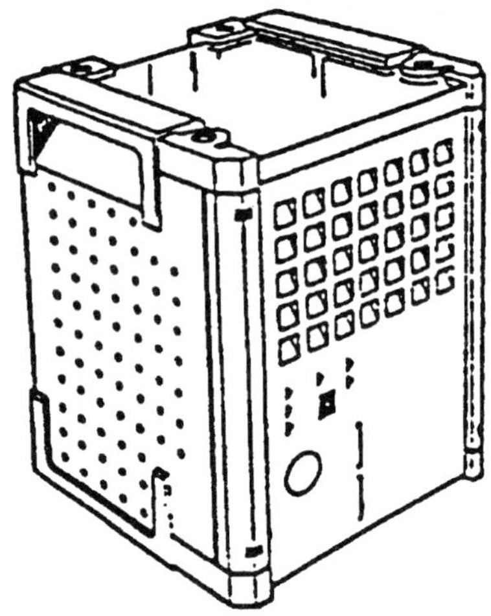

*Figure 9.2 Inverter for a stand-alone system (Mastervolt, Amsterdam).*

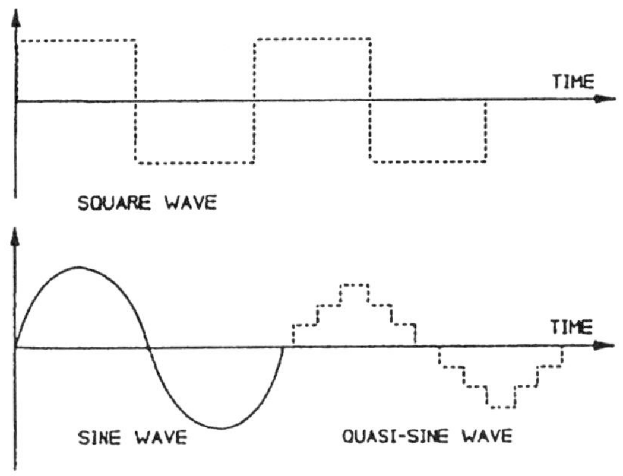


*Figure 9.3 Inverter waveforms.*

The third waveform shown in Figure 9.3 is quasi-sine wave or modified sine wave. This waveform has multiple steps and closely approximates a true sine wave. Modified sine wave is the form of power that modern stand-alone inverters produce and that is acceptable to many AC appliances. Modern inverters are from 85% to 95% efficient and draw minuscule amounts of power in standby mode (for example: 1 W, wake up: 10 W).

In stand-alone applications sizing of the inverter is very critical. The unit must be large enough to handle motor-starting surge inrush currents and the resultant short-duration peak loads. However, care must be taken to avoid oversizing the unit because it will not deliver its peak efficiency when operated at only a fraction of its rated power (see Figure 9.4).

Ideally an inverter for a stand-alone photovoltaic system should have the following features:

- surge capacity (2...4 times $P_n$),
- low idling and no-load losses,
- output voltage regulation,
- low battery voltage disconnect,
- low harmonics,
- high efficiency,
- low audio and RF-noise.

## 9.3 Grid-connected inverter

### 9.3.1 General

With a photovoltaic array on the rooftop of his house a home owner can produce electric energy for his residential loads (Figure 9.5).
Grid connected photovoltaic plants become part of the utility system. The essential device of a grid interactive photovoltaic installation is the inverter. It acts as an interface between the solar array and the utility grid.

The utility-interactive inverter differs from the stand-alone unit in that it can function only when tied to the utility grid. This inverter converts direct current produced by the solar cells into 'utility grade' alternating current that can be fed into the distribution network. The utility-interactive inverter not only conditions the power output of the photovoltaic array, it also serves as the system's control and the means through which the site-generated electricity enters the utility lines. It uses the prevailing line-voltage frequency on the utility line as a control parameter to ensure that the PV system's output is fully synchronized with the utility power.

The overall system performance depends heavily on PCU performance. The waveform of the inverter output current should be of almost perfect sine wave shape.

The static power inverter includes a possible means for controlling the entire photovoltaic system. This includes sensing the available array power and closing a grid (AC) side contactor to begin operating as soon as possible after sunrise. At night the inverter should be completely switched off.

The control logic of the inverter should include a protection system to detect abnormal operation conditions such as:

- earth fault, DC-side;
- abnormal utility conditions (line voltage, frequency, loss of a single phase);
- inverter switch off when the power stage is overheating.

The inverter should be protected against transient voltages with varistors on the DC- and on the AC-side. The power available from the solar array varies with module temperature and solar insolation. The inverter has to extract the maximum power out of the solar array. Therefore it is equipped with a device called 'Maximum Power Point Tracker' (MPPT-unit). With the help of the MPPT-unit the inverter input stage varies the input voltage until the maximum power point (MPP) on the array's IV-curve is found. A new MPP should be searched at least every 1 to 3 minutes.

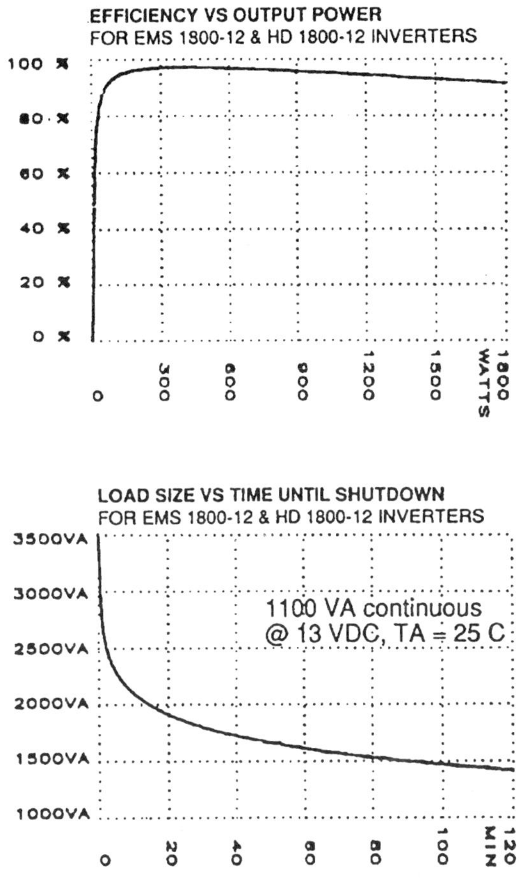


*Figure 9.4 Efficiency and overload capability of a quasi-sine wave inverter.*

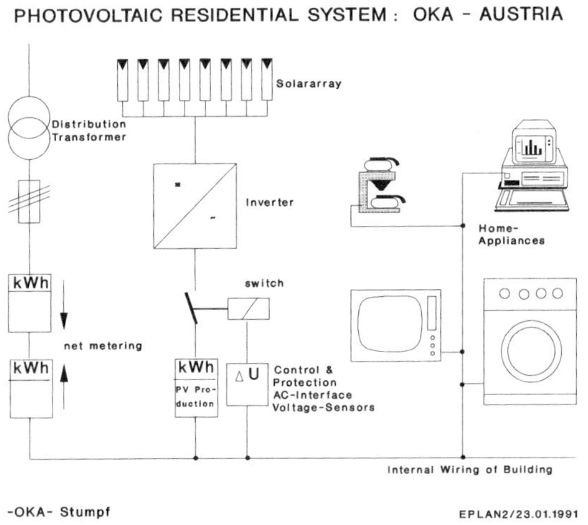


*Figure 9.5 Utility-interactive PV system.*

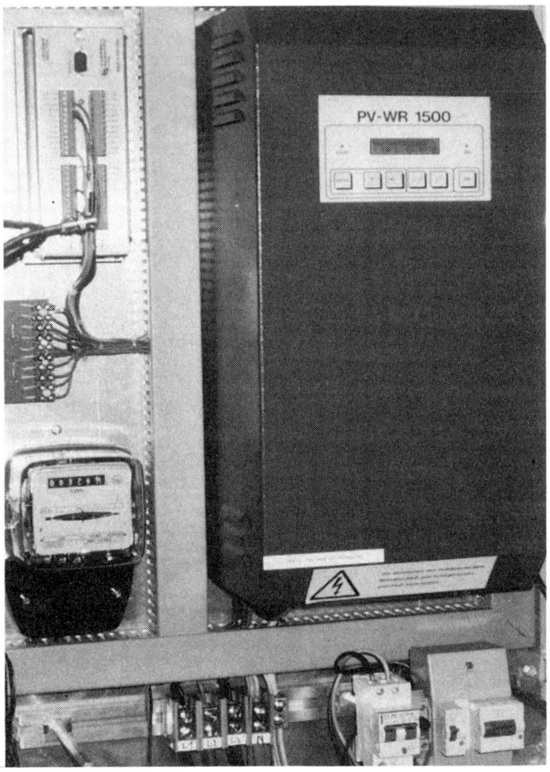


*Figure 9.6 Utility-interactive inverter, 1.5 kW.*

### 9.3.2 Inverter specific data

Basic information to be obtained from the inverter manufacturer or dealer:

- Cost (including any required options),
- Array compatibility: number of modules per string, power tracking range,
- Utility compatibility: power quality, harmonics, power factor, electrical isolation, islanding prevention,
- Energy performance: weighted-average efficiency, no load losses, standby losses,
- Warranty provisions,
- Maintenance and repair.

Site information needed for inverter selection:

- PV system size: $kW_{peak}$,
- Electrical environment: DC system voltage, local safety code requirements, single-phase or 3-phase system,
- Physical environment: type of location, humidity, dust, temperature, noise,
- Utility connection requirements: safety, power quality, trip limits, protection details.

Recommended inverter specifications:

- High conversion efficiency > 92% for $P/P_n > 0.1$,
- Low start-up and shut-down thresholds,
- Power factor > 0.85 (satisfies local utility requirements),
- Low total harmonic distortion of output current: k < 3% at full power (EN 60555),
- Maximum power point operation,
- No shut down if the array power exceeds rated power: -> current limiting function,
- Low power consumption at night: $P_o < 0.5\%$ of $P_n$,
- Automatic disconnect at utility fault conditions (deviations of V, f),
- Automatic restart after fault is cleared,
- AC-ripple of array voltage < 3%,
- Low level of audible noise,
- Low level of RF-emissions measured on AC- and DC-side, VDE 871 B (1.1.1996),
- Type of cooling, e.g. fan,
- Electric isolation between DC- and AC-side,
- Overvoltage protection at DC- and AC-side,
- High availability.

### 9.3.3 Line-commutated inverter

The traditional line-commutated inverter (see Figure 9.7) is commonly used in drive units for induction motors. The power stage is equipped with thyristors (SCR, silicon-controlled rectifier). For solar applications the control unit has to be modified to allow for MPP-tracking. Furthermore the driver circuit has to be changed to shift the firing angle from the rectifier mode ($0° < \alpha < 90°$) to the inverter mode ($90° < \alpha < 180°$). In Europe this type of solar inverter is available with a rated power of 1.5 kVA as a single phase unit. Three-phase devices are installed in larger PV systems. Up to a rated power of 300 kVA six pulse units are used (SMA, PV-Neurather See, RWE). Twelve pulse inverters are a little more expensive but do not produce so much harmonics (two times 450 kVA, PV-Toledo/RWE&Endesa&Union Electrica Fenosa).

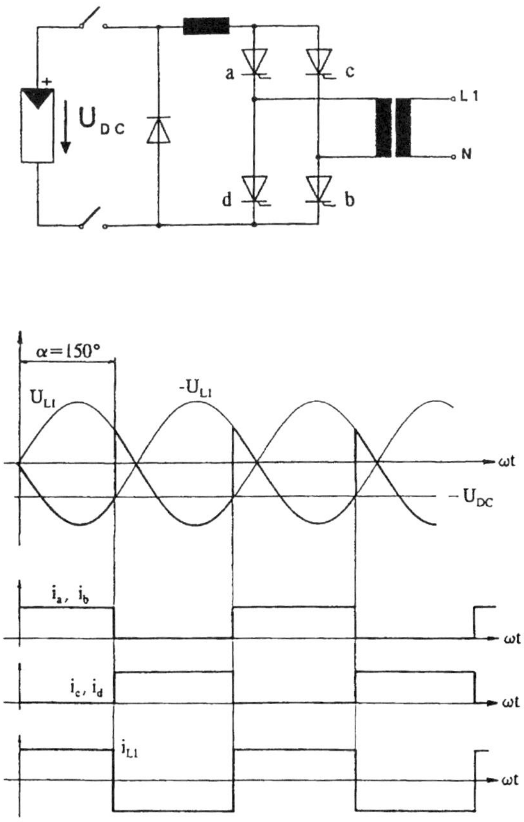


*Figure 9.7 Line-commutated inverter.*

Thyristor type inverters need a low impedance grid interface point for commutation purposes. A line-commutated inverter should not be used if the maximum power available at this grid extension is less than five times the rated power of the inverter. The advantage of the line-commutated inverter is its low price (0.6 to 1 US $/W). The disadvantage of this system is the poor power quality of the AC electricity. The harmonic content of the current fed into the grid is quite large (Figure 9.8). Without additional filter circuits the limits of the European Standard EN 60555 will be exceeded. Another disadvantage is the poor power factor which is measured to be 0.6 to 0.7 inductive (Figure 9.10). External phase shift equipment has to be installed to meet the utilities' power factor requirements (better than 0.9). For small systems (<10 kW) the best choice is to use a pulse width modulated inverter unit with MOSFET power transistors or an IGBT power stage.

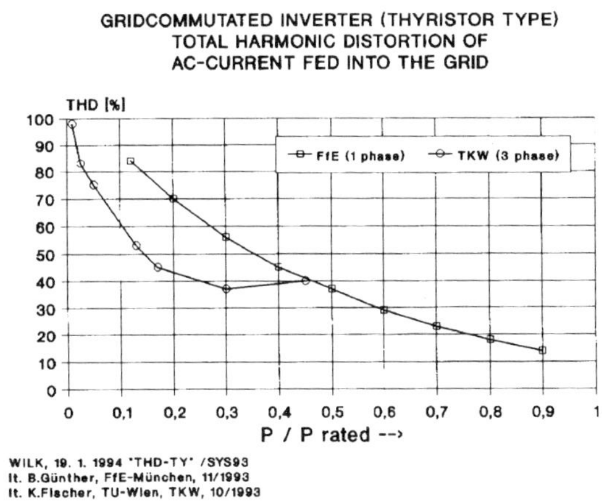


*Figure 9.8 Total harmonic distortion of AC current, line-commutated inverter.*

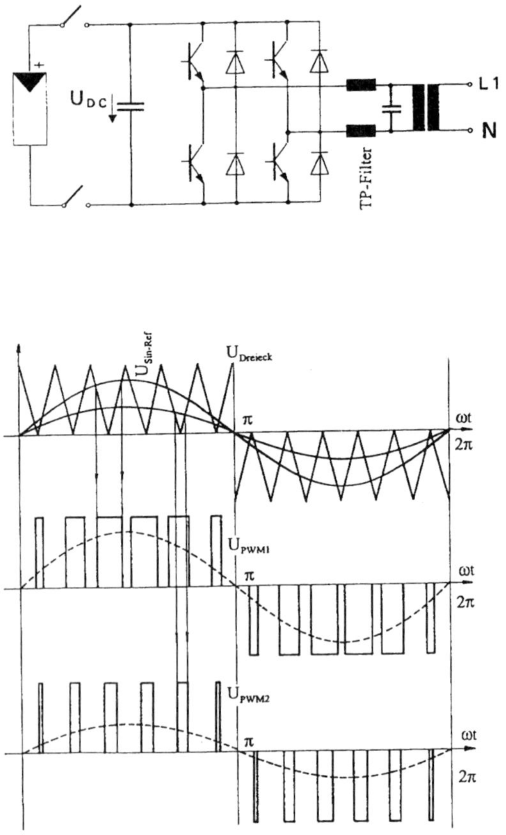


*Figure 9.9 Self-commutated inverter with pulse width modulation and low frequency transformer.*

### 9.3.4 Self-commutated inverter

With pulse width modulation (PWM) an almost perfect sine wave is shaped (THD<1%). Figure 9.9 provides the information on how pulses of different length and polarity are combined in order to form a sine wave. Lower harmonic distortion can be achieved by using higher switching frequencies. But switching losses will rise with higher frequencies and overall efficiency will be lower. Several different semiconductor devices can be used in the power stage of a self-commutated inverter:

- MOS transistors,
- IGBTs (isolated gate bipolar transistors) or
- GTO thyristors (gate turn off thyristors).

MOS transistors are used in units up to 5 kVA. They have the advantage of low switching losses at higher frequencies.

GTOs are used in very large installations (100 kVA and more). Because of the limited switching frequency the shape of the sine wave is not perfect and additional filtering is needed.

The isolated gate bipolar transistor (IGBT) is composed of a power pnp-transistor and a n-channel MOS-transistor in Darlington configuration. For turning on the device a voltage has to be applied at the gate. The required driving power is very low (mW to W) and depends on the input capacity of the gate and the switching frequency. Because the on-state voltage drop is 2 V DC, the system voltage should be higher than 200 V. With the presently available IBGTs inverters up to 200 kVA can be built without paralleling the power switches. The power stages of IGBT inverters have an uncomplicated structure. IGBTs are readily available in half-bridge configuration.

**Power Factor**

Self-commutated inverters can be easily connected to the grid (self-commutated, line-synchronized) presenting the advantage of near unity power factor, not loading the grid with reactive power or requiring large power factor compensation networks (Figure 9.10).

**Input Ripple**

The inverter is operating in a switching mode and current pulses are drawn from the DC source (PV array). Therefore, the inverter's input voltage is not constant. A ripple can be measured upon the array voltage. A too large ripple reduces the power output of the system because the array cannot be operated in the maximum power point of the IVcurve.

**Overload Capability**

The inverter must be able to limit the output current to a safe level if the input is overloaded by the solar array. This happens in special weather conditions with irradiance values of more then 1000 W/m² (snow reflections, multiple reflections on white clouds). The inverter varies the input voltage to higher values and reduces the output power of the array by leaving the maximum power point. Most system designers decide to use inverters with power limiting function. Therefore,

$$1.2 < P_{ARRAY}/P_{INVERTER} < 1.4$$

is a good choice. Some line-commutated inverter units switch off at overload conditions. In this case the ratio of $P_{ARRAY}/P_{INVERTER}$ should not exceed the value of 1.

### 9.3.5 Solar inverter with high frequency transformer section

The low frequency (50/60 Hz) transformer of a standard inverter with pulse width modulation is a very heavy and bulky component. When using frequencies of more than 20 kHz, a ferrite core transformer is the best choice. For a 1.8 kVA inverter this type of transformer is the size of a fist and is a lightweight component.

The inverter consists of five sections (Figure 9.11):

a) width modulation
b) high frequency inverter with pulse width modulation
c) high frequency rectifier
d) low pass filter
e) output stage.

This type of inverter needs two additional stages compared to the inverter concept in Figure 9.9 (Standard PWM inverter). Because of the high switching frequencies a lot of filters are needed at the input and output side of this inverter to avoid radio frequency emissions. In general electrical household appliances have to operate within the limits of VDE 871 B (1.1.1996). Up to now most inverters of this type are only specified according to VDE 871 A.

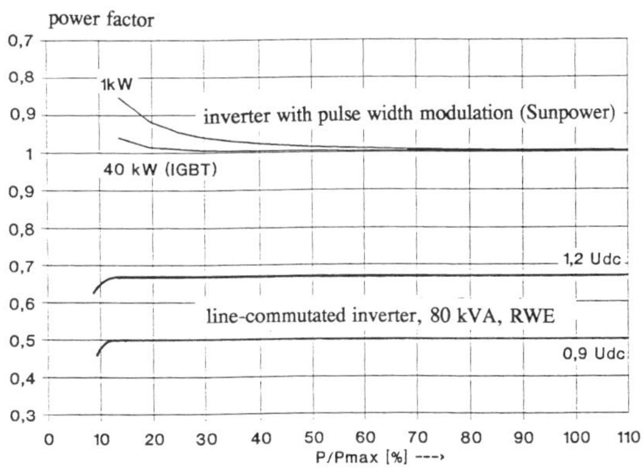


*Figure 9.10 Power factor of different inverters.*

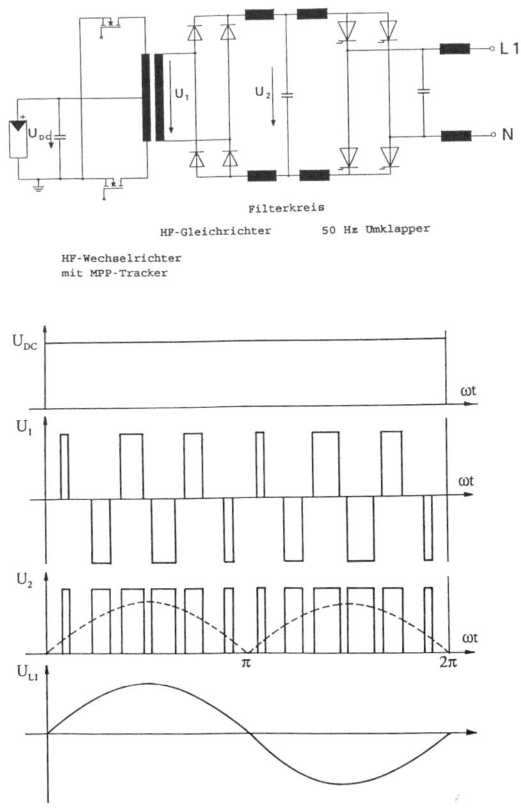


*Figure 9.11 Inverter with high frequency transformer section (SMA PV WR 1800).*

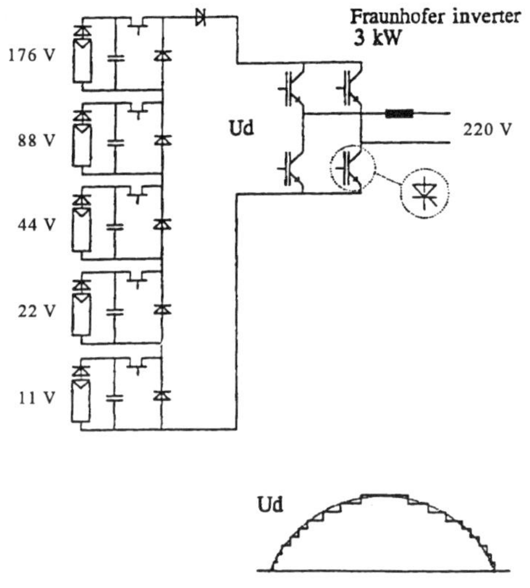


*Figure 9.12 The FhG-ISE Inverter.*

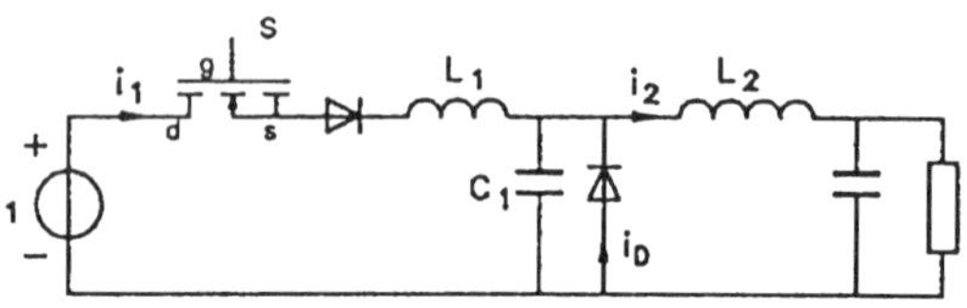


Principle step down chopper with Zero Current Switching.

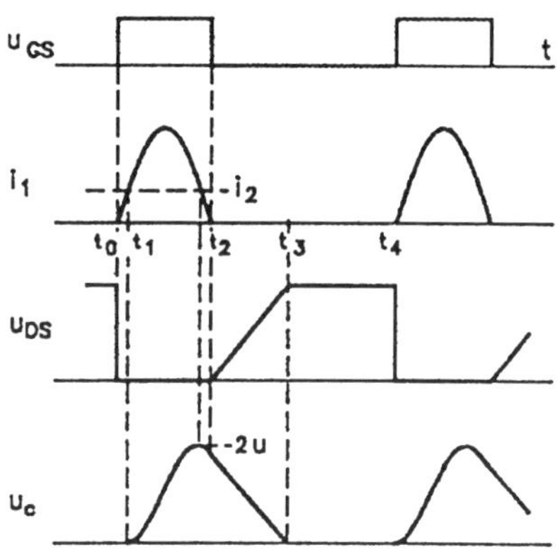


Waveforms of ZCS step down chopper

*Figure 9.13 Zero Current Switching Principle (ECN Petten).*

### 9.3.6 Transformerless inverter with binary switching concept

This type of inverter was especially designed by FhG-ISE for photovoltaic applications. The principle is based on using five different PV arrays (Figure 9.12). The array voltages are e.g. 11 V, 22 V, 44 V, 88 V and 176 V. The voltage values are ordered according to the binary system. Triggered by a sine-wave generator and fast high-power switches a 230 V AC wave form can be synthesized with 5 bit accuracy.

The electronic switches connect as many arrays in series as necessary to follow the shape of the grid's voltage continuously. Some of the unique features of this inverter are:

- high efficiency,
- low no load losses (7 W/10 kW),
- low harmonic content,
- low weight, small size.

Due to the special working principle, the solar generator has to be wired in a more complex way. Five times more cables and lightning protection devices are needed for this special design. There is no maximum power point tracking device included. Depending on the weather conditions not all of the five arrays are loaded with the optimal voltage.

A similar working principle can also be applied to stand-alone applications.

### 9.3.7 Inverter with resonant switching stage

The traditional power conditioning units are often referred to as 'hard-switching' inverters. The term hard-switching is used because the current flow is interrupted by changing the conductance of the power switch (this is done via a control signal on the control input of the power device). During switching the voltage over the power switch increases and after just a short time (100 ns to several ms) the current flow decreases rapidly. During this time the dissipation in the power switch is relatively high, while the very fast current and voltage transients can cause **E**lectro **M**agnetic **I**nterference problems (EMI).

The term 'soft-switching' inverter is used for high frequency units based on the principle of resonant switching technology. This technique emerged for power supply technology in recent years. Due to the resonance phenomenon the current becomes zero at certain times (depending on the resonance frequency). The resonance effect is initiated by opening or closing the electronic switch. If it is switched off exactly when the current is zero, the switching loss will be very low (Zero Current Switching - ZCS). It is also possible to switch off when the voltage is zero (Zero Voltage Switching - ZVS). Because of the low switching losses this operating principle allows very high frequencies (10 kHz to 10 MHz). Because of the high switching frequency all capacitive and inductive components can be small and relatively cheap.

In Figure 9.13 the basic principle of a ZCS quasi resonant DC converter is shown. Due to the use of new magnetic material for the high frequency (RF) transformer the physical size of that component can be reduced (100 W→size: 3x3x2 $cm^3$, while for a normal 50/60 Hz toroidal transformer the diameter is 10 cm and the height is about 5 cm).

### 9.3.8 Module-integrated converter

In grid-connected operation the PV generator is connected to the utility grid via a DC/AC converter. In a standard configuration the PV generator consists of a number of parallel strings which in turn consist of solar modules connected in series.

Depending on the system size the string voltage in grid connected systems reaches values from about 50 V up to over 700 V. In such a system a lot of DC cabling is required carrying a relatively high current, which causes some losses. PV installations with a central inverter have shown problems with respect to high DC voltage levels, risk of DC arcs, fire hazard and protection. Most of these problems can be overcome by a careful system lay-out, special cabling and DC switches/fuses, which increases the cost at system level.

The integration of PV module and inverter into so-called 'AC-Modules' (also referred to as MIC, **M**odule **I**ntegrated **C**onverter ) offers interesting possibilities to overcome the problems mentioned above. The module integrated converters generate floating AC power at module level. The size of such an inverter is indicated in Figure 9.14.

The AC modules do justice to the inherent modularity of photovoltaic building blocks. A PV system can be connected simply at 230 V AC level with standard low cost AC installation techniques. For AC modules both hard and soft-switching inverters are applied.

The block diagram for a typical AC module is shown in Figure 9.15.

*Figure 9.14 Module Integrated Converter (110W, Mastervolt 130).*

Because of the specific structure and location of AC modules, several aspects are a matter of concern:

- because of the location of the AC modules the converters are operating under severe climatic conditions;
- the inverter of an AC module should be very reliable, because of difficult access and a higher number of devices installed (MIC: 110 W vs. Standard: 1.800 W);
- islanding protection has to be realized in each unit or a central device must be installed to guarantee safe operation (info bus cabling);
- earth leakage detection is necessary in each AC module inverter.

AC modules have a potential to realize PV systems in a cost-effective way. System design with high modularity can be achieved combined with reduced installation cost and lower cabling losses. Up to now the principle has been demonstrated by several companies. Field experience has to be gained yet.

## 9.4 Inverter costs

Up to now the expenses for solar modules tend to dominate system costs of residential PV installations (50%). The power conditioning unit of many grid-connected PV applications costs less than 13% of the total expenses. As inverters are electronic solid state devices, they have a high cost reduction potential. By starting mass production (MIC), the price could be reduced rather rapidly. With the 'Photovoltaic Rooftop Programmes' organized in several countries solar inverters are produced in quantities of several hundred to more than thousand.

Prices already came down from 6500 US $/kW in 1985 to 1100 US $/kW in 1994. Thyristor type inverter units with more than 10 kW rated power are now available for 600 US $/kW (SMA, 1994).

## 9.5 Inverter reliability

Plant availability is determined very much by the performance of the inverter. Module failures are not very often the reason for PV system outages. Overload conditions often occur at alpine sites. One thyristor type inverter automatically stops operation and starts again the next morning. Many inverters adapt to DC-overload with a current limiting function. Some devices are very sensitive to overvoltage and undervoltage conditions on the utility grid. Overvoltages can be measured at the end of long grid extensions when only few loads are connected. This situation is quite common on a sunny Sunday afternoon in a rural area grid.

When operating pilot plants with prototype inverters, it should be possible to reach availabilities of more then 95% after the first months of testing and optimizing. Normally the repair activities are completed within several hours. Most energy losses are caused by the time it takes to get the spare parts from the inverter company Small inverters are easier to send to the company for repair.

If the PV system is hit by a lightning strike the inverter will almost certainly be damaged.

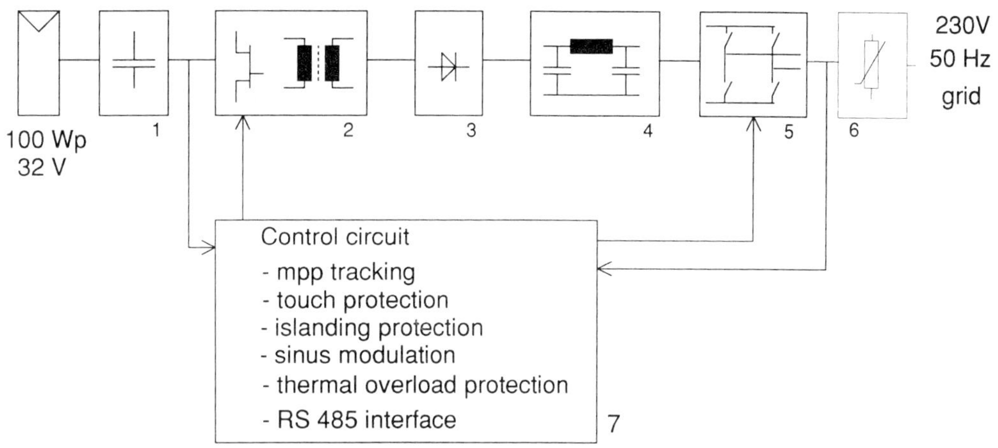


*Figure 9.15 Block diagram AC module (OKE/ECN Petten).*

## 9.6 Integration of PV systems into the utility network (Inverter - utility interface)

### 9.6.1 General

Grid-interactive photovoltaic systems are co-operating with the utility network. Since the interconnection involves the two-way flow of energy, each side has responsibilities. The PV system must incorporate features that ensure safety and quality of the utility service. This results in specific requirements for the power conditioning hardware (Figure 9.16). When the photovoltaic system is interconnected with the utility distribution network a two-way flow of electric energy will be established (Figure 9.17).

**Solar Fraction**

Only a certain fraction of the PV electricity can be used in the appliances of the residence at the same time it is generated. The utility grid will absorb excess photovoltaic generated energy during midday hours when residential energy usage is relatively low (Figure 9.18). During bad weather conditions and at night-time the utility generators will supply electricity to the residential loads (back-up). The solar fraction depends on the size of the solar array and the load curve of the house and is presented in Figure 9.19. If the PV system is very small almost all produced PV electricity will be consumed in the loads of the residence. The larger the PV array, the more PV electricity is fed into the distribution network of the power company.

**Metering**

Most utilities in Europe have adopted net metering (Figure 9.17). Two seperate, reverse blocked meters for sold and purchased electric energy are installed. To get precise information on the amount of PV electricity produced by the solar electric system, an additional meter is recommended (meter 1, Figure 9.17).

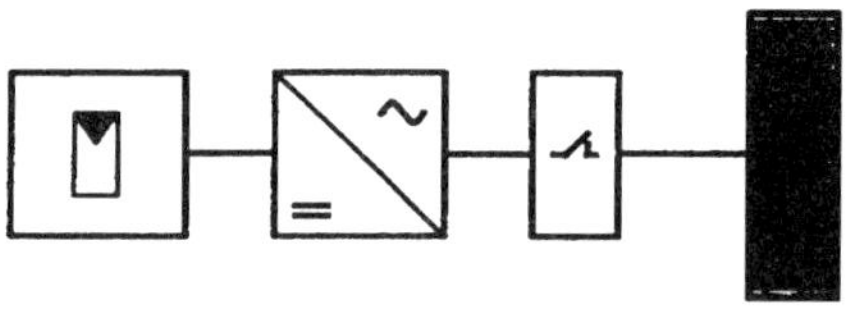

*Figure 9.16 Grid-interactive PV system.*

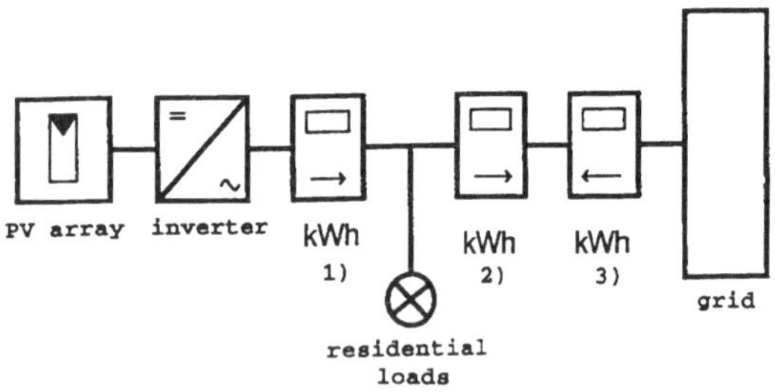


*Figure 9.17 Metering, residential PV system.*

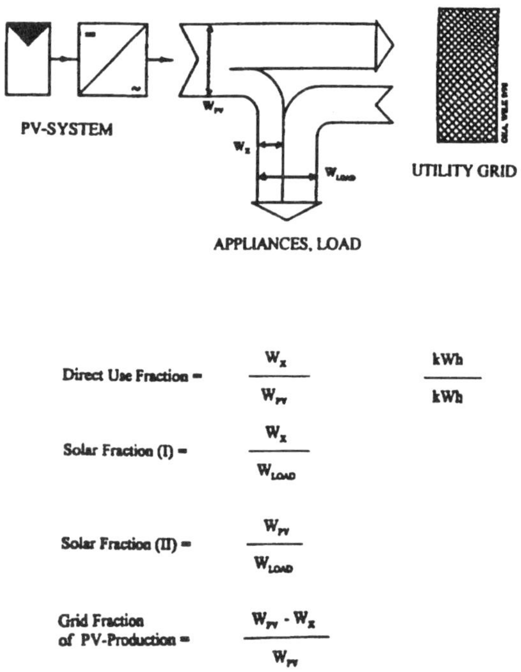


*Figure 9.18 Energy flow between PV system, utility grid and household loads.*

Meter 2 (Figure 9.17) is a three-phase unit that counts the amount of electricity sold to the power company. Meter 3 is the standard three-phase instrument which is used to measure the electric energy purchased from the power company. Meter 2 will not be needed if the PV system is operated in a country where a buy-sell ratio of unity is applied. In this case meter 3 will count up and down (Switzerland).

### 9.6.2 Utility Interface Requirements

For the utility company linemen safety and power quality is a major concern. Utility staff are used to having a one-way flow of electricity in most parts of their distribution network. When they are switching off one feeder at the substation, they do not expect that a dead line will be energized by a PV system. With the installation of dispersed PV generating units in the distribution network the utilities have to modify their traditional manner of operating the grid. Furthermore the operational differences between an autoproducer with a rotating generator and a PV inverter have to be accepted.

Key issues:

- PV system becomes part of the utility system;
- inverter must satisfy utility grid quality requirements;
- linemen's safety is a major concern;
- PV systems should never energize a 'locked out' or dead line;
- automatic disconnect of inverter at utility fault conditions;
- lockable outdoor disconnect switch (accessible to utility personnel);
- operating at essentially unity power factor;
- electrical isolation between PV system and grid.

Utilities in European countries have adopted different safety concepts at the PV/utility interface. In most cases the inverter has to be disconnected within five seconds when a utility fault condition occurs. The safety concepts for a single-phase and a three-phase inverter PV installation are described in Figures 9.20 and 9.21.

Although most inverters are not able to operate without the grid voltage (islanding) many utilities only connect a PV system to the grid if a three-phase voltage relay is installed. The inverter must be disconnected from the grid by the relay in case the output voltage exceeds or falls below predefined limits. The recommended range for the voltage tolerance is 80-110% of nominal voltage (Germany, Spain, Italy, Austria). All three phases must be monitored to be able to detect the loss of the grid voltage. Even if a single phase inverter can keep the voltage stable on one phase (islanding) the voltage relay will detect the loss of the voltage of the remaining two phases and switches off the whole unit. In Austria an external tree-phase relay must be used. In Germany the internal control unit of the inverter has to do the monitoring of all three phases. In Switzerland the inverter is connected to one phase and no relay has to be installed. In Germany and in Austria an outdoor disconnect switch is needed when three-phase inverters are used. The main switch has to be accessible to utility personnel. Whenever the inverter output exceeds any of the pre-defined conditions during operation (over/under voltage, over/under frequency) the PV plant has to be disconnected automatically from the grid. Reconnection to the grid is to be attempted only after a certain time delay (three minutes) to enable grid control systems to attempt fault correction. The overall intention of these measures is to prevent the inverters from harmful influences from the grid as well as to protect the grid network, including all loads, from inverter failures (see also IEA PV Power Systems Implementing Agreement Task V).

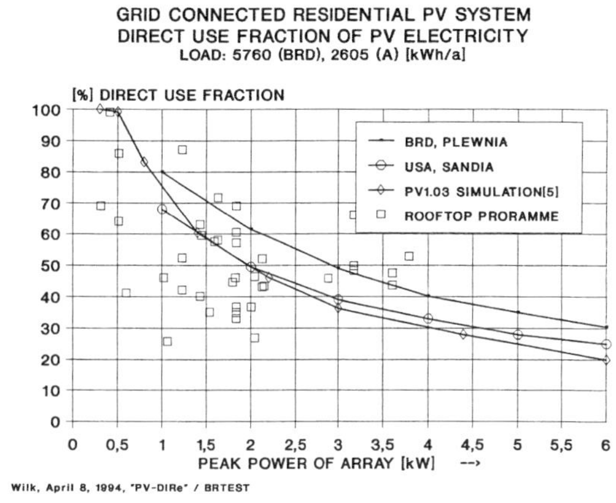


*Figure 9.19 Annual average fraction of PV system output coincident with on-site loads.*

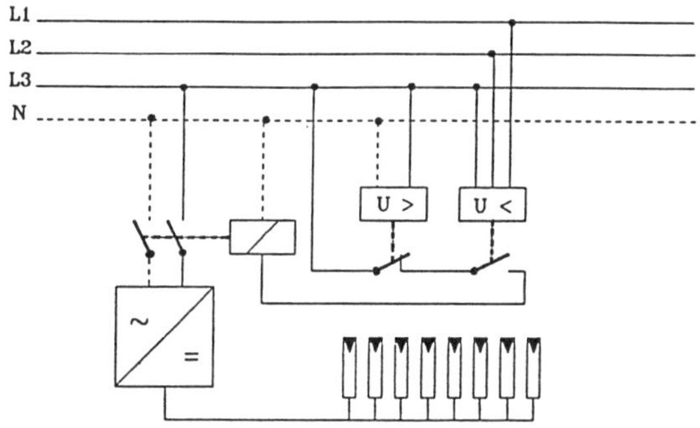


*Figure 9.20 Safety concept for a single-phase inverter PV installation.*

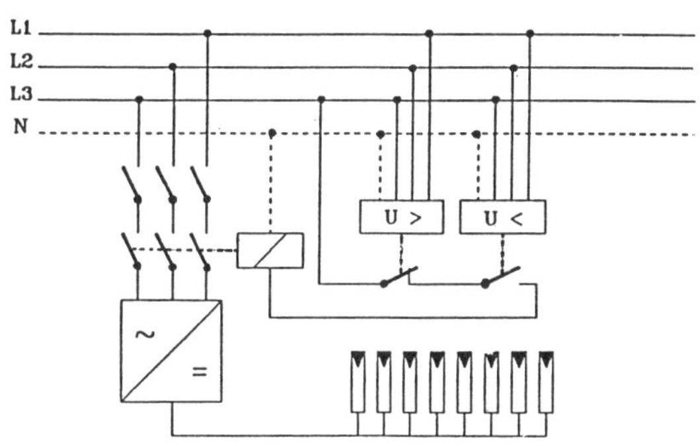


*Figure 9.21 Safety concept for a three-phase inverter PV installation.*

### 9.6.3 Inverter specifications related to utility requirements

The inverter must detect utility fault conditions and disconnect itself immediately within five seconds. 'Islanding' must be prevented under all circumstances.

Fault conditions are:

- over/under voltage (e.g. 80% - 110%)
- over/under frequency (50/60 Hz +/- 1%)
- loss of one phase.

In Germany inverters have to be designed to monitor the voltages of all three lines of the grid. This regulation applies to one-phase and three-phase inverters. With this feature included grid outages should be detected and 'islanding' can be avoided easily. In Switzerland single phase units can be connected to the grid without three-phase monitoring. In this case 'islanding' has to be prevented by other means (inverter software etc.). Austrian utilities rely on three-phase monitoring with an over/under voltage relay that must not be part of the inverter (similar to Spain, Italy). The price of a protection relay equals approximately 0.5% of the total system cost. The relay can easily be checked with a small test box.

The advantage of this method is that utility personnel do not have to check many different types of inverters but have to check only the standard relay. After the fault is cleared, the inverter has to restart operation again automatically.

## 9.7 The value of PV electricity

The value of PV generated electricity is judged very differently from country to country. Different buy-sell ratios can be found. In most countries the value of PV electricity will be higher if almost all solar electricity can be used in the home owners appliances. If the principle of 'avoided costs' is used, the value of excess energy will depend on the utility's generating costs. In many countries generating costs vary with time of day and underlie seasonal variations. A combination of different generating units is used by the utilities with different actual costs of generation. As several utilities purchase electricity from autoproducers at the rates of 'avoided costs' identical tariffs are often used for PV systems and small hydro power plants. In countries with a special PV funding system the buy-sell ratio becomes almost unity (e.g. Germany).

Several countries have special tariffs for PV autoproducers that include 'external costs' of traditional electricity production. Some communities decided to pay even higher rates for electric energy generated in PV plants, like

- Burgdorf, Switzerland SFr 1.- / x kWh
- Freising, Germany DM 2.- / x kWh.

Further discussions in Switzerland will eventually lead to the application of the principle of 'marginal costs'. The value of PV electricity would then be approximately 0.3 SFr/kWh. This is the price of 1 kWh that is produced by a newly-built conventional power plant.

## 9.8 Efficiency issues, electric yield

The electric output of a grid-connected PV system depends heavily on the performance of the inverter. As all the precious solar electricity passes through the power conditioner to be fed into the grid, inverter efficiency is a very important quality. In Central Europe's climatic conditions PV plants are operated at rated power only for several hundred hours a year.

Overall system performance depends heavily on the inverter characteristics:

- stand-by losses: power consumed at night-time by the micro processor control unit (1 - 17 $W_{AC}$ for a 2-kW system);
- self consumption: $P_0$: 0.5 to 4% of rated power, DC input power needed to start inverter operation (for driver circuits, magnetizing the transformer core etc.);
- maximum efficiency: 88 to 94% for 2-kW inverters.

Stand-by losses will be accumulated over 4380 night-time hours. This energy flows from the grid to the inverter unit. Self consumption reduces the available DC input power during approximately 4380 operating hours of the plant (daytime, Central Europe).

Self consumption influences the monthly values of inverter efficiency. Because of many hours of part-load conditions in wintertime, the inverter efficiency will then be lower (Figure 9.23).

To optimize the annual electric output of a PV plant it is essential to match the array size and the rated power of the inverter. The peak power of the solar modules should not be lower than the rated power of the inverter to avoid part-load operation. If the array is too large, energy will be lost because of the current limiting function at inverter overload conditions (see Figure 9.22).

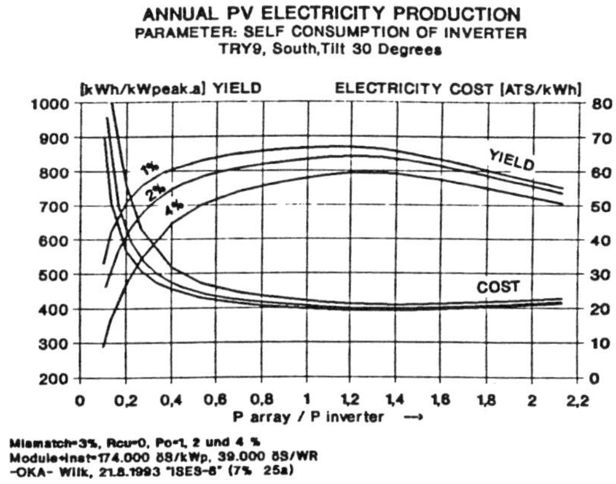


*Figure 9.22 Annual electricity output of grid-connected PV plant in Munich/Germany.*

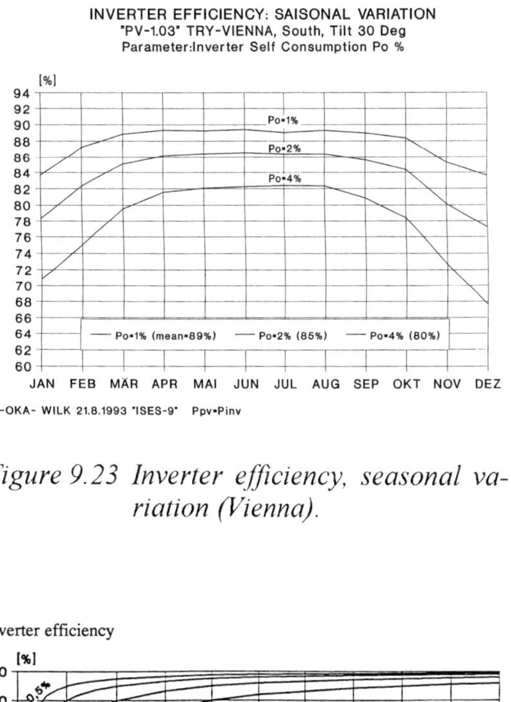


*Figure 9.23 Inverter efficiency, seasonal variation (Vienna).*

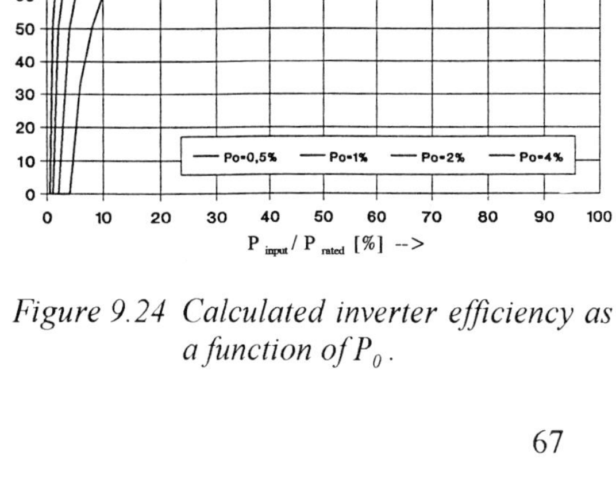


*Figure 9.24 Calculated inverter efficiency as a function of $P_0$.*

## Chapter 10

# Hybrid Power Systems

### 10.1 Introduction

Electrical energy requirements for many remote applications are too great to allow the cost-effective use of stand-alone or autonomous PV systems. In such cases, it may prove more feasible to combine several different types of power sources to form what is known as a 'hybrid' system. To date, PV has been effectively combined with other types of power generators such as wind, hydro, thermo-electric, petroleum-fueled and even hydrogen. The selection process for hybrid power source types at a given site can include a combination of many factors including site topography, seasonal availability of energy sources, cost of source implementation, cost of energy storage and delivery, total site energy requirements, etc.

Examples for Hybrid PV/Generator systems are shown in Figure 10.1 and 10.4.

### 10.2 Petroleum-fueled engine generators (gensets)

Petroleum-fueled gensets (operating continuously in many cases) are presently the most common method of supplying power at sites remote from the utility grid such as villages, lodges, resorts, cottages and a variety of industrial sites including telecommunications, mining and logging camps, and military and other government-operated locations.

Although gensets are relatively inexpensive in initial cost, they are not inexpensive to operate. Costs for fuel and maintenance can increase exponentially when these needs must be met in a remote location. Environmental factors such as noise, carbon oxide emissions, transport and storage of fuel must also be considered.

*Figure 10.1 Hybrid PV/Generator System Example. Courtesy Photron Canada Inc., Location: Sheep Mountain Interpretive Centre, Parks Canada Kluone National Park, Yukon Territories, Canada, 63° North Latitude; Components shown include: generator (120/240 V), battery (deep cycle industrial rated @ ± 10 kWh capacity), DC to AC stand-alone inverter (2500 W @ 120 V output), miscellaneous safety + control equipment including PV array disconnect, PV control/regulator, automatic generator start/-stop control, DC/AC system metering etc.; components not shown: PV array (800 W peak).*

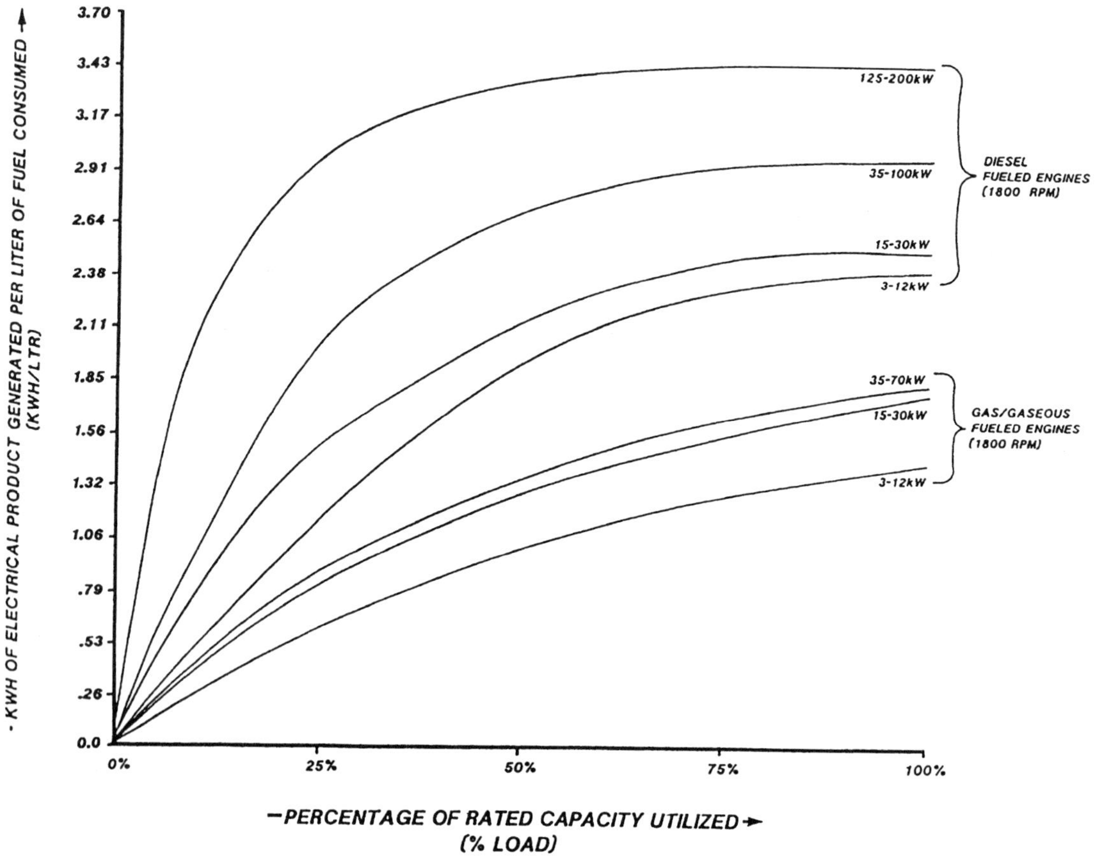


*Figure 10.2 Genset fuel efficiency vs. capacity utilized (Photron Canada Inc.).*

Fuel to power conversion efficiencies may be as high as 25% (for a diesel-fueled unit operating at rated capacity). Under part load conditions, however, efficiencies may decline to a few percent. Considerable waste heat is therefore available and may be utilized for other requirements such as space and/or water heating.

Figure 10.2 shows the genset fuel efficiency as a function of the capacity utilized.

## 10.3 Why a PV/genset hybrid?

PV and genset systems do not have much in common. It is precisely for this reason that they can be mated to form a hybrid system that goes far in overcoming the drawbacks to each technology. Table 10.1 lists the respective advantages and disadvantages.

As the sun is a variable energy source, PV system designs are increased in size (and therefore cost) to allow for a degree of system autonomy. Autonomy is required to allow for provision of reliable power during 'worst case' situations, which are usually periods of adverse weather, seasonally low solar insolation values or an unpredicted increased demand for power. The addition of autonomy to the system is accomplished by increasing the size of the PV array and its requisite energy storage system (the battery).

When a genset is added, additional battery charging and direct AC load supply capabilities are provided. The need to build in system autonomy is therefore greatly reduced. When energy demands cannot be met by the PV portion of the system for any reason, the genset is brought on line to provide the required backup power. Substantial cost savings can be achieved and overall system reliability is enhanced.

PV/genset hybrid systems have been utilized at sites with daily energy requirements ranging from as low as 1 kWh per day to as high as 1 MWh per day, which illustrates their extreme flexibility. They are a proven and reliable method for efficient and cost-effective power supply at remote sites.

## 10.4 PV/genset hybrid system description

The PV/genset hybrid utilizes two diverse energy sources to power a site's loads (see Figure 10.3). The PV array is employed to generate DC energy that is consumed by any existing DC loads, with the balance (if any) being used to charge the system's DC energy storage battery. The PV array is automatically on line and feeding power into the system whenever solar insolation is available and continues to produce system power during daylight hours until its rate of production exceeds what all existing DC loads and the storage battery can absorb. Should this occur, the array is inhibited by the system controller from feeding any further energy into the loads or battery.

A genset is employed to generate AC energy that is consumed by any existing AC loads, with the balance (if any) being used by the battery charger to generate DC energy that is used in the identical fashion to that described for the PV array above.

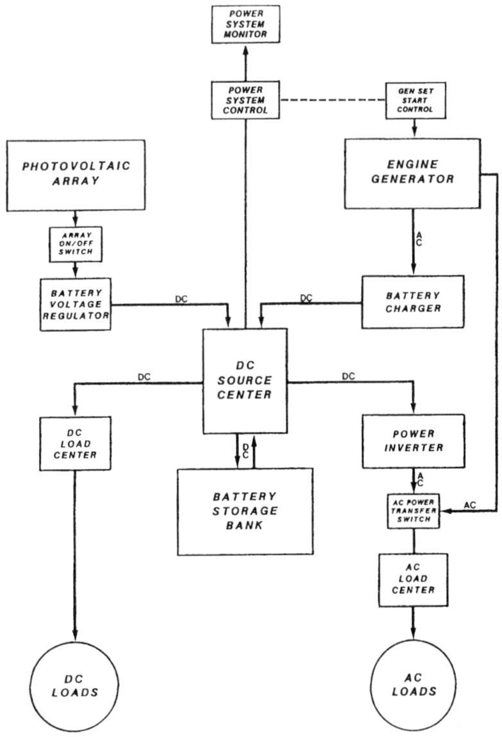


*Figure 10.3 Block diagram of a hybrid PV-Genset system (Photron Canada Inc.).*

At times when the genset is not running, all site AC power is derived from the system's power conditioner or inverter, which automatically converts system DC energy into AC energy whenever AC loads are being operated.

The genset is operated cyclically in direct response to the need for maintaining a suitable state of charge level in the system's battery storage bank.

## 10.5 Other PV/hybrid types

*Figure 10.4 Hybrid PV/Generator System Example. Courtesy Photron Inc., Location: Caples Lake, California, USA; 65 kVA 3 ⌀ @ 480 V generator which includes co-generation equipment (i.e. heat exchangers to utilize the thermal energy created by unit operation).*

Certain specific site locations may offer access to other forms of power generation. Access to flowing water presents the potential for hydro power. Access to consistent wind at sufficient velocity presents the potential for wind power. PV/hydro and PV/wind hybrid systems have been utilized at sites with daily energy requirement ranges similar to those described for PV/genset hybrids. Their use, however, is much more site dependent, as their energy source is a factor of that location's topography.

PV/thermoelectric generator hybrid systems have been used effectively at sites whose daily energy requirement is relatively low, ranging from 1 to 20 kWh per day. Propane is the fuel source for the thermoelectric process, and conversion efficiencies of up to 8% can be achieved. Considerable waste heat is therefore available which may be utilized for other requirements. In cold climates, this heat is often used to maintain the battery storage system at desired temperature levels.

| | **Advantages** | **Disadvantages** |
|---|---|---|
| **Genset** | Low initial expense<br>On-demand power<br>High power density<br>Widely known<br>Highly portable | High operating cost<br>High maintenance<br>Non-renewable fuel<br>Noise pollution<br>Air pollution |
| **PV** | Renewable fuel<br>Reliable/low maintenance<br>Versatile/modular<br>Non-polluter | Low power density<br>High initial expense<br>Sunshine dependent<br>Not widely known |

*Table 10.1 Relative Advantages of Energy Sources: Genset vs. PV.*

### Section C

# ARCHITECTURAL INTEGRATION

## Principal Contributors

**Chapter 11:** **Introduction to Architecture and Photovoltaics**

Ingo Hagemann (Planungbüro Hagemann, Germany)
Steven Strong (Solar Design Assiciates, USA)

**Chapter 12:** **Photovoltaic Modules Suitable for Building Integration**

Peter Toggweiler (PMS, Switzerland)

**Chapter 13:** **Design Concepts**

Cinzia Abbate-Gardner (Consultant to ENEL, Italy)
Peter Toggweiler (PMS, Switzerland)
Gregory Kiss (Kiss and Company, Architects, NY, USA)
Tony Schoen (ECOFYS, Netherlands)
Heinz Hullmann (Inst. f. d. Industrialisierung des Bauens, Germany)

**Chapter 14:** **Integration Techniques and Examples**

Peter Toggweiler (PMS, Switzerland)
Hermann Laukamp (Fraunhofer ISE, Germany)
Christian Roecker (EPFL-LESO, Lausanne)

## Chapter 11

# Introduction to Architecture and Photovoltaics

### 11.1 Motivation

The last two decades have brought significant changes to the design profession. In the wake of traumatic escalations in energy prices, shortages, embargoes and war along with heightened concerns over pollution, environmental degradation and resource depletion, awareness of the environmental impact of our work as design professionals has dramatically increased.

In the process, the shortcomings of yesterday's buildings have also become increasingly clear: inefficient electrical and climate conditioning systems squander great amounts of energy. Combustion of fossil fuels on-site and at power plants add greenhouse gases, acid rain and other pollutants to the environment. Inside, many building materials, furnishings and finishes give off toxic by-products contributing to indoor air pollution. Poorly designed lighting and ventilation systems can induce headaches and fatigue.

Architects with vision have come to understand it is no longer the goal of good design to simply create a building that is aesthetically pleasing - buildings of the future must be environmentally responsive as well. They have responded by specifying increased levels of thermal insulation, healthier interiors, higher-efficiency lighting, better glazing and HVAC (heating, ventilation and air conditioning) equipment, air-to-air heat exchangers and heat-recovery ventilation systems. Significant advances have been made and this progress is a very important first step in the right direction.

However, it is not enough. For the developed countries to continue to enjoy the comforts of the late twentieth century and for the developing world to ever hope to attain them, sustainability must become the cornerstone of our design philosophy. Rather than merely using less non-renewable fuels and creating less pollution, we must come to design sustainable buildings that rely on renewable resources to produce some or all of their own energy and create no pollution.

One of the most promising renewable energy technologies is photovoltaics. Photovoltaics (PV) is a truly elegant means of producing electricity on site, directly from the sun, without concern for energy supply or environmental harm. These solid-state devices simply make electricity out of sunlight, silently, with no maintenance, no pollution and no depletion of materials. Photovoltaics are also exceedingly versatile - the same technology that can power water pumping, grain grinding, communications and village electrification in the developing world can produce electricity for the buildings and distribution grids of the industrialized countries.

There is a growing consensus that distributed photovoltaic systems which provide electricity at the point of use will be the first to reach widespread commercialization. Chief among these distributed applications are PV power systems for individual buildings.

Interest in the building integration of photovoltaics, where the PV elements actually become an integral part of the building, often serving as the exterior weathering skin, is growing world-wide. PV specialists from some 15 countries are working within the International Energy Agency's Task 16 on a five-year effort to optimize these systems and architects are now beginning to explore innovative ways of incorporating solar electricity into their building designs.

This chapter presents the reasons behind building-integrated PV and some examples of this early work. View these PV-powered buildings as a first glimpse into the coming new era of energy-producing buildings where this elegant, life-affirming technology will become an integral part of the built environment.

*Figure 11.1 SOS Kinderdorf, Zwickau, Germany: 2.9 $kW_p$ roof-integrated PV system. Frameless architectural laminated glass with amorphous silicon cells.*

## 11.2 Building envelope

The building envelope is composed out of the roof, the facade and the parts which have a contact to the ground. It is the interface between the outside and inside world. Especially the roof and the facade have to face changing and also often unpredictable demands. Facade materials, building components and construction techniques that are visible on the inside and outside of the building have to fulfill numerous requirements:

***Building envelope Performance Requirements***

***Weather condition requirements***

- *Protection against rain;*
- *Resistance to moisture;*
- *Protection against ice, hail and snow;*
- *Frost resistance;*
- *Be a heatshield;*
- *Allow heat adsorption and heat storage;*
- *Protection against sun;*
- *Allow admission of light and light conditioning;*
- *Control light diffusion;*
- *Sunlight reflexion, passage and control;*
- *Allow passage of wind for cooling in summer;*
- *Protection against cold winds in cold season;*
- *Avoid wind noises;*
- *Protection against glare.*

***Structural requirements***

- *Provide structural stability;*
- *Allow load-carrying capacity;*
- *Resist internal and external loads;*
- *Protection against risk of mechanical and chemical damage;*
- *Protection of building elements against coming-off;*
- *Avoid condensation;*

- *Easy maintenance or maintenance-free;*
- *Be durable;*
- *Be vandal-proof;*
- *Provide fire protection;*
- *Allow emergency exit.*

***Occupant requirements***
- *Allow incidence of light;*
- *Allow contact and communication with the outside;*
- *Protection against view inside the building (privacy);*
- *Allow spatial separation;*
- *Allow passage.*

***Town planning and design requirements***
- *Allow dialogue with urban surrounding;*
- *Give a spatial image;*
- *Appearance;*
- *Provide design possibilities;*
- *Choose appropriate materials, textures and forms;*
- *Be representative;*
- *Serve corporate identity.*

***Emissions requirements***
- *Keep heat emission to a minimum;*
- *Keep noise inside the building;*
- *Avoid passage of odours to outside.*

***Enclosure requirements***
- *Protection against waste gas from the urban surrounding;*
- *Provide radiation shielding;*
- *Avoid odour nuisance;*
- *Keep out noises;*
- *Avoid insects and vermin infiltration;*
- *Keep out dust;*
- *Keep out pollen.*

## 11.3 Planning context of an energy conscious design project

The possibilities of an active and passive solar energy use in buildings is greatly influenced by the form, design and construction of the building envelope.

***Active and Passive Solar Design Principles***

***I. Heating Strategies for Buildings***
- *Collection of Solar Gains in Cold Climate;*
- *Storage and Restorage of Solar Heat Gains;*
- *Distribution of Solar Heat Gains;*
- *Avoidance of Heat Losses;*
- *Use Active Solar Systems like Hot Water Collectors or Photovoltaics;*
- *Use Hybrid Systems.*

***II. Cooling Strategies***
- *Control of Solar Radiation;*
- *Reduction of External Solar Heat Gains;*
- *Reduction of Internal Heat Gains;*
- *Natural Cooling and Ventilation Techniques.*

***III. Use of Daylighting Techniques to Reduce Cooling Loads and Energy Demands***

A promising potential of active solar energy use is the production of electricity with photovoltaics. This technology can be adapted to existing buildings as well as to new buildings.

It can be integrated into the roof, into the facade or into different building components, such as a photovoltaic rooftile. Such an integration makes sense for various reasons:

- The solar irradiation is a distributed energy source; the energy demand is distributed as well.

- The building envelopes supply sufficient area for PV generators and therefore
- Additional land use is avoided as well as costs for mounting structures and energy transport.

In order to use PV together with other available techniques of active and passive solar energy, it must be considered that some techniques fit well together and others exclude each other.

For example: As a kind of a 'passive cooling system', creepers are used for covering the south facade of a building. The leaves evaporate water and provide shade on the facade. This helps to avoid penetration of direct sunlight and reduces the temperature in the rooms behind the facade.

But at the same time these leaves also create shading on PV modules that may be mounted on the facade resulting in a far lower electricity production (see Figure 11.3).

To avoid such design faults it is necessary to compare and evaluate the different techniques that are available for creating an energy conscious building.

Therefore in the future an overall energy concept for a building must be made already at the beginning of the design process. The architect and the other experts involved in the design and planning process need to work together right from the beginning of the design and planning process. Together they have to search right from the beginning for the best design for a building project (see Figure 11.4).

*Figure 11.2 Church, Steckborn, Switzerland: 18.6 $kW_p$ facade-integrated PV system. The monocrystalline PV modules cause a creative upgrading of the church tower (see also colour plate 3).*

*Figure 11.3 HEW, Hamburg, Germany: 16.8 $kW_p$ facade-integrated PV system. The polycrystalline PV modules are installed as fixed shading devices.*

## 11.4 Photovoltaics and architecture

Photovoltaics and architecture are a challenge for a new generation of buildings.
PV systems will become a modern building unit, integrated into the design of roofs and facades. To create an outstanding overall design a close collaboration between engineers and architects is essential.
The architect, together with the engineers involved are asked to integrate the PV at least on four levels during the planning and realization of the building:

- Design of a building (shape, size, orientation, colour)
- Mechanical integration (multi-functionality of a PV element)
- Electrical integration (grid connection and/or direct use of the power)
- Maintenance and operation control of the PV system must be integrated into the usual building maintenance and control.

The photographes illustrate some examples of PV house designs.

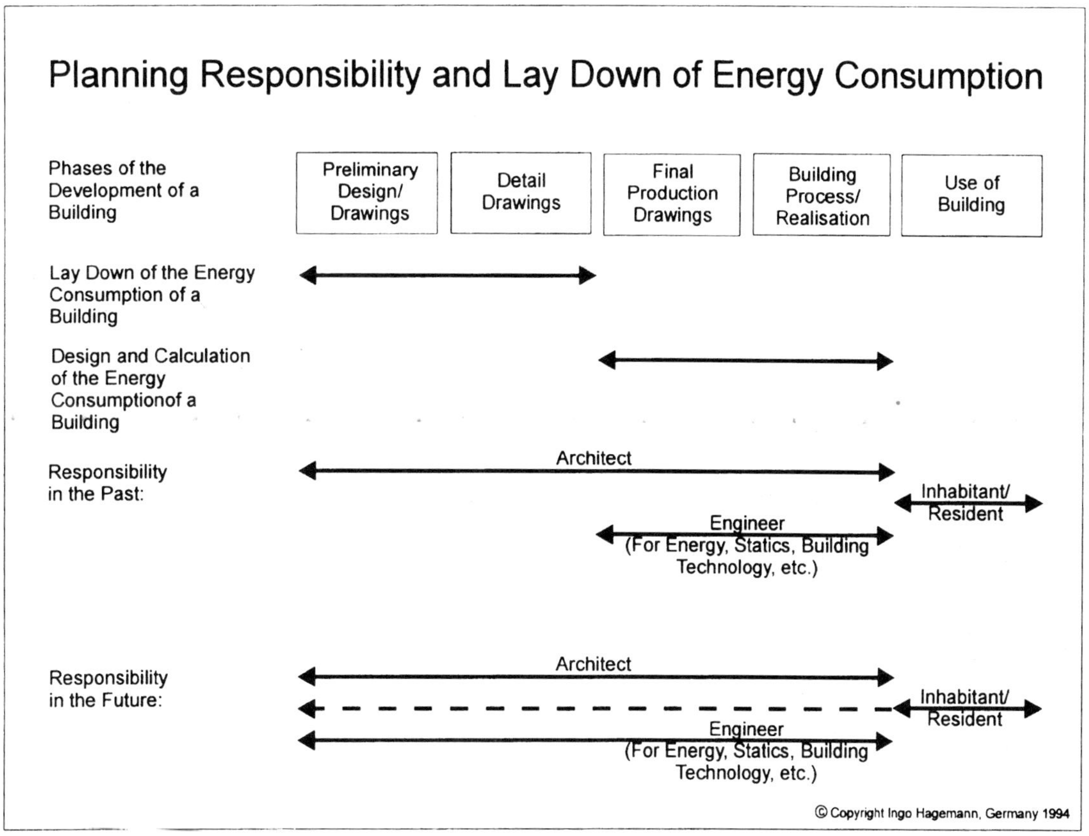


*Figure 11.4 Planning Responsibilities and Lay Down of Energy Consumption.*

## Chapter 12

# Photovoltaic Modules Suitable for Building Integration

### 12.1 Introduction

A PV module is basically designed and manufactured for outdoor use. All products available are suitable for exposure to sun, rain and other climatic influences. These circumstances make possible the use of PV modules as part of the building skin. Many different types of module technologies are available. However, not all of them are useful for building integration. In the past, modules used to be specifically designed for energy generation and the building element function was neglected. Fut-ure considerations aim at designing building elements which also deliver electrical energy. The module configuration is described in Chapter 5. The various module types such as monocrystalline, polycrystalline and amorphous silicon have differing aesthetic considerations. The colour of monocrystalline cells varies from uniform black to a dark grey with a uniform surface structure. In contrast, the structure of polycrystalline cells shows irregular grey-to-blue coloured crystals. In both types, the current gathering grid lines are well visible as a silver or black metallic colour. This may change in the future, as the grid colour is better matched to the colour of the cells. Many types of modules and laminates are available with mono- or polycrystalline cells. For semi-transparent modules the space between the single cells is enlarged to let light pass through. Custom-designed modules allow an individual quality of the light transmission. In addition, the colour of the back sheet for non-transparent modules can be selected.

Amorphous silicon is deposited on metal, glass or plastic films, meaning different kinds of modules are available. These modules usually have a dark brown colour. For semi-transparent modules of amorphous silicon the cells themselves are pervious to light. Since the cells absorb a part of the spectrum, the colour of the passed light is changed.

Today's technology of module design has led to several solutions for building-integrated PV systems. The following table shows the advantages and disadvantages of different types of PV modules.

*Figure 12.1 Standard laminate (APS).*

| | Module construction technique | Typical dimension [cm²] | Application suitability | | | | |
|---|---|---|---|---|---|---|---|
| | | | Sloped roof | Flat roof | Wall | Window | Shading |
| 12.1 | Standard laminates as above without frames | 33 x 130<br>45 x 100<br>55 x 115 | + | + | + | - | + |
| 12.2 | Standard modules with plastic or metal frame (glass multi-layer non-transparent back sheet) | 33 x 130<br>45 x 100<br>55 x 115 | + | o | o | - | o |
| 12.3 | Roofing modules (tiles/slates) | to fit with standard roofing systems | + | - | - | - | o |
| 12.4 | Glass-glass modules with predefined transparency | all dimensions between 15 and 200 | o | o | + | + | + |
| 12.5 | Glass modules with transparent plastic back sheet (predefined transparency possible) | all dimensions between 15 and 200 | o | o | + | + | + |
| 12.6 | Modules with metal back sheet and plastic cover | 15 x 150 | + | + | + | - | + |
| 12.7 | Custom-designed modules | various dimensions | + | + | + | + | + |

\+ = high suitability
o = low suitability
\- = not suitable

*Table 12.1 Suitability of different module types for building integration (see also to the corresponding Figures 12.1-12.7).*

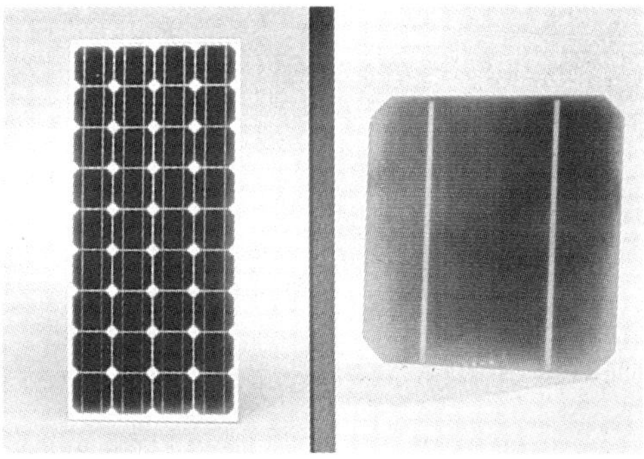

Figure 12.2 Standard module (BP Solar).

Figure 12.3 Swiss solar tile (Newtec).

Figure 12.4 Large glass-glass module with predefined transparency (Flagsol).

*Figure 12.5 Glass module with predefined transparency (Sanyo).*

*Figure 12.6 Module with metal back sheet and plastic cover (USSC).*

*Figure 12.7 Custom-designed module (here: triangular shape) (Solution).*

## Chapter 13

# Design Concepts

### 13.1 Design as multidisciplinary approach

The design and construction of buildings has shifted away from the creation of a 'dwelling', moving towards the interpretation of the multifunctionality of the various construction typologies, as well as the components of architecture. The tendency to industrialize building construction has revealed the need to clarify the design process, which controls the growing complexity of the project, and the need to combine the technical-scientific role of the engineer with the technical and composite role of the architect, and with the new industrial technologies. The problem of architecturally integrating photovoltaic technology requires an interdisciplinary design approach. This not only imposes collaboration and the presence of highly specialized professionals on the project team, but also introduces a sensitivity to problems that go beyond the building itself. These inhabit a sphere that is even broader, including social, economic, environmental, energetic and ecological issues.

As an example, a building facade must not only keep out water and regulate heat loss, it must also regulate the entry of light, provide a sound barrier, offer ease of technical maintenance and also must be aesthetically and architecturally satisfying.

The study of the elements and components involved in photovoltaic application systems is of great importance in the final quality of the product. On this level, a PV building installation is a typical industrial design problem. Each project for the architectural integration of photovoltaics may require both the revision of the type and the dimensions of the photovoltaic panel, and the study of new framing and mounting systems. Individuals possessing the skills that are specific to this sector will be involved as well. These people range from the architects and engineers to photovoltaic cells manufacturers and to all the technicians and the producers of the materials employed in the construction.

### 13.2 The role of photovoltaics in building

It has been mentioned before that a building is a combination of many complex systems: structural, mechanical, electrical, and others. Changes to the parameters of one system affect the others. An assessment of the performance of a building-integrated PV system as an element of a building skin therefore requires a multidisciplinary approach. A building-integrated PV system in fact adds another role to this programme: power generation. A building surface can usually be classified as a roof or a wall, with significant differences in function, construction, and thermal and solar loading.

Architects are turning into industrial designers and project energy management consultants in order to address PV construction as more than just a building material. The value of PV can be significantly enhanced by collateral energy benefits. As PV cells capture sunlight and convert it to electricity, they also convert it to heat which can be used or discarded as necessary.

## 13.3 Considerations in designing building envelopes for PV

During the design stage, both the technical and aesthetic characteristics of the PV module must be considered in order to arrive at a satisfactory integration of PV into the building as a whole.

Central to the study of PV building design is the conflict between PV solar considerations and contemporary building conventions. The primary goal in the layout of PV power systems is to maximize the amount of power generated through optimum array orientation, but this goal is tempered in the case of building design by considerations of construction costs, optimum building floor area, daylight control, thermal performance, and aesthetics. In addition, building envelopes are often designed to deflect and minimize the amount of radiation falling on a building's surface, since cooling is often the largest consumer of energy in a building. By contrast, photovoltaics require the greatest possible radiation in order to maximize performance. Also, the envelope system must be designed to resist any water which may permeate the skin and infiltrate not only the framework, but also the photovoltaic panel interlayer. In addition, electrical connections which penetrate the weather seal must also be designed for reliable performance. In climates where clear/cold days bring substantial and immediate temperature changes to building surfaces, the PV envelope system needs to resist or eliminate the condensation.

Both mild and extreme climates require good insulating properties at the envelope. PV panels may be directly laminated with insulation or may be incorporated into multi-layer air- or gas-filled insulating units. Electrical connection design will also need to take into account thermal bridging.

The impact of lightning on PV building envelopes is another important environmental issue. PV structures will need to be grounded and circuited to prevent a possible power surge which may result in damage to the panels or a hazard to their occupants.

The issues of building-integrated PV design are not exclusively technical. The balance between the issues of PV building design and construction will vary greatly according to the circumstances of each project (climate, budget, client priorities, aesthetics, etc.).

## 13.4 Outline for the design procedure

The following discussion attempts to identify a possible outline for the design procedure of a building-integrated PV system.

### 13.4.1 Climatic considerations and orientation

Locations, climates, latitude, average cloudiness, average temperatures, precipitation, humidity, dust/dirt, wind loads, and seismic conditions will all affect the economics of a PV-integrated building system by virtue of how they are addressed in the envelope design.

South-facing unobstructed PV panels are usually oriented at a tilt equivalent to the local latitude in order to receive maximum solar radiation, while building walls are generally vertical for reasons of economy, efficiency and traditional construction technology. East and west-facing facades perform relatively well at steep angles or vertical orientation. They still yield 60% of optimally inclined south oriented PV systems due to the low angle of the sun at the beginning and end of the day. The building's electrical load profile and the utilility time-of-use rate structure are important parameters in establishing optimum PV orientation.

The inherent flexibility of PV compared to other types of solar collectors (wiring is inherently easier to run than plumbing), combined with anticipated low material cost (thin-film PV on glass substrates are basically similar to coated architectural glass) raises the possibility of considering PV a building material first, and a PV device second. Thus, an architect desiring a monolithic appearance for a building may choose to clad all of a building's surfaces with PV, even those that will never see the sun. If the economics permit, PV can be used in any number of building component configurations and without an overriding demand for optimal orientation. If PV orientation is not perceived as a design restriction, architects will be much more open to their use.

### 13.4.2 The site

The characteristics of the site and its orientation will influence the design in order to determine where and how to integrate PV in the building.

Building floor area can be a precious commodity. In some cases, sloped PV panel configurations will reduce the amount of occupiable perimeter floor area because the wall effectively 'cuts back' on floor area as the building gets taller. Any reduction in usable floor area needs to be considered when evaluating the life-cycle costs of a PV system.

High rise structures are usually built in an urban environment where real estate costs are high and the surrounding landscape is dense. Shadows cast by other tall buildings reduce the performance of the panels. It may be that for certain high rise projects only the upper stories will be clad in PV with only some areas active during the course of a day. In such cases, the designer can chose to articulate or camouflage the active and inactive portions of the facade by using contrasting or matching cladding next to the PV.

### 13.4.3 Zoning regulations and building codes

Sometimes zoning regulations regarding colours, building shape, and building codes, will also have an impact on the selection of the different kind of photovoltaic material to use.

Another parameter is certainly the identification of technology, or which type of panels (single crystal or polycrystalline cells) should be used. This decision should be made in accordance with the dimension of the building, its possible overall configuration, the economic constraints (some technologies are more expensive than others) and the energy demand.

### 13.4.4 Type of panels

The selection of the PV panels, their aesthetic characteristics in terms of modular geometry, dimension, colour, mounting system (with exposed frame or without frame), will influence the overall appearance of the building and the architectural character of the intervention. The modules are the most immediately recognizable and distinguishing components of the various PV systems. They are most visible from outside the building and will most probably be placed in a prominent position to avoid the shadow cones of nearby buildings that could reduce the efficiency of the system and the desired electric energy output. PV panels may provide energy benefits beyond the electricity they generate by providing passive solar heating or cooling load reduction.

The modular geometry, the colours and the texture of both the cells and the glass of the photovoltaic panels constitute the main aesthetic characteristics of the panels. The balance between the amount and quality of glass and the quantity and type of cells used in the panel is part of the design process and will be relevant to both the panel and the space. This device can be used to exploit a sort of decorative and functional texture formed by the natural light that filters through the cells. In fact, this transparent grid of light created by the glass space between the cells is a significant contribution to the enrichment of the architectural quality of the indoor spaces. Very different are those cases in which the panel is interpreted as real construction material. Here it acquires considerable functional and aesthetic validity, particularly when we analyze the very interesting dialectic taking place between PV and traditional construction materials.

The photovoltaic cell presents its most interesting aesthetic aspects on the exterior face which is exposed to the sun. By varying the selection and positioning of PV cells in the module, it is possible to obtain a wide variety of colour, brilliance, reflectivity and transparency effects.

The cables, the junction box and the battery of a PV module might be visible, depending upon the type of installation and the type of module. This should be taken into consideration during the design stage, so that the overall spatial configuration is controlled.

### 13.4.5 Installation

For both new construction and retrofit, the method of installation is important to the cost effectiveness of the system. For example, glazing installation from the interior does not require exterior scaffolding. Interior glazing is a common method of contemporary curtain wall installation today, accomplished by splitting the mullion and muntin extrusions into separate elements which snap into place in the field.

Shop labour is usually cheaper and more precise than field labour. Whenever possible, panelized or prefabricated wall or roof sections will save money, especially in complex PV wall profiles. Prefabricated systems could also include some electrical balance of systems or PV-powered devices such as fans or lights.

It is important to recognize that the integrated nature of applied architectural PV installations (curtain wall framing, glazing systems, PV and electrical connections, etc.) will require the combined efforts of a number of different building trades and jurisdictions. Conventional construction sequences and responsibilities must be considered in the development of PV products if they are to fit present construction industry practice.

### 13.4.6 Structure, engineering and details

The mechanical and electrical systems required to maintain and operate a building of substantial scale are often complex and can require a tremendous amount of energy. Photovoltaics add both benefits and additional levels of complexity to the engineer's task. PV buildings will challenge architects and engineers to develop innovative solutions for integrating a building's support system with PV supplied power. Critical issues to consider for engineering systems integration will be, among others, safety, durability and economy.

## 13.5 Configurations for PV building integration

Before starting to design a PV building, it is very important to analyze each possible applicable solution of application to determine its overall impact on the building's energy balance and the energy efficiency performance of the overall system.

The following diagrams offer a panorama of configurations for PV integration, selected by three main architectural application typologies:

- walls and facades,
- roofs and large coverings,
- light filtration and screening elements.

### 13.5.1 Walls and facades (Figures 13.1 - 13.7)

There are two basic curtain wall framing systems in common use: pressure plate and structural silicone glazing. In pressure plate systems, the glazing unit is mechanically held from the front by a plate with an extruded cover or 'cap'. Structural silicon glazing glues some or all of the glazing edges to the framing systems. In pressure plate systems, the mullion cap depth must be kept to a minimum to avoid adverse shadowing on PV cells. Alternatively, flush application of a structural silicon seal between PV glazing units eliminates shadowing effects but increases weather seal and durability problems for PV panel edges.

To minimize sealing problems, or to capture heat from the PV modules, it may make sense to fabricate a double wall envelope, where the PV glazing is the external, unsealed layer and the inner layer may be the weather tight enclosure.

Optimizing PV panel performance in building wall applications may require complex detailing and therefore higher construction costs in order to accommodate optimal orientations to the sun. For wall applications, these complex configurations may take on a sloped or 'sawtooth' profile (see Figures 13.3, 13.4, 13.13) or the PV may be applied independently of the building's skin as awnings or light-shelves (Figures 13.14, 13.15).

### 13.5.2 Roofs and large coverings (Figures 13.8 - 13.13)

For roof applications, installations generally require little compromise in solar orientation but may create structural or weather-proofing problems. Horizontal roof configurations must be structured to accommodate other types of loading such as snow and water. This issue requires different solutions depending upon the location of the building. In climates with considerable snow accumulation, skylights and roof systems incorporating PV may need to be designed with a slope sufficient to shed snow which may be steeper than the optimal slope for solar gain. Partial PV skylight enclosures will shade interior spaces from direct sunlight while simultaneously harnessing power from the sun's rays. PV roof monitors could also reduce or eliminate the need for daytime electric lighting by providing indirect daylight.

Flexible substrates such as sheet metal or fabric would open up large new markets. PV devices that mimic traditional building materials (such as clay roof tiles) may also increase their market.

### 13.5.3 Light-filtration and screening elements (Figures 13.14, 13.15)

Opaque PV panels installed as PV window awnings (Figure 13.14) or PV light shelves (Figure 13.15) can shield direct sun while providing diffuse, indirect light to interior spaces. The portion of these light shelves which are exposed to sunlight would be PV; the portion in shade could be any reflective material. The panels' surface would bounce light onto the ceiling inside.

Another PV device with some passive solar benefits is the semi-transparent photovoltaic panel, or PV window, designed to admit a specific amount of light and/or view to a space. Some thin-film PV devices are inherently semi-transparent, if produced with clear conductive coatings on glass substrates. Alternatively, opaque PV devices may be rendered effectively transparent by the creation of a pattern of clear areas where the opaque materials have been removed. It should be possible to incorporate semi-transparent PV into insulating or high-performance multi-pane glazing units. With a less active PV area, the solar performance of these semi-transmissive panels will be less than with opaque PV. But the passive benefits and vision area produced in some cases will outweigh the reduction in efficiency.

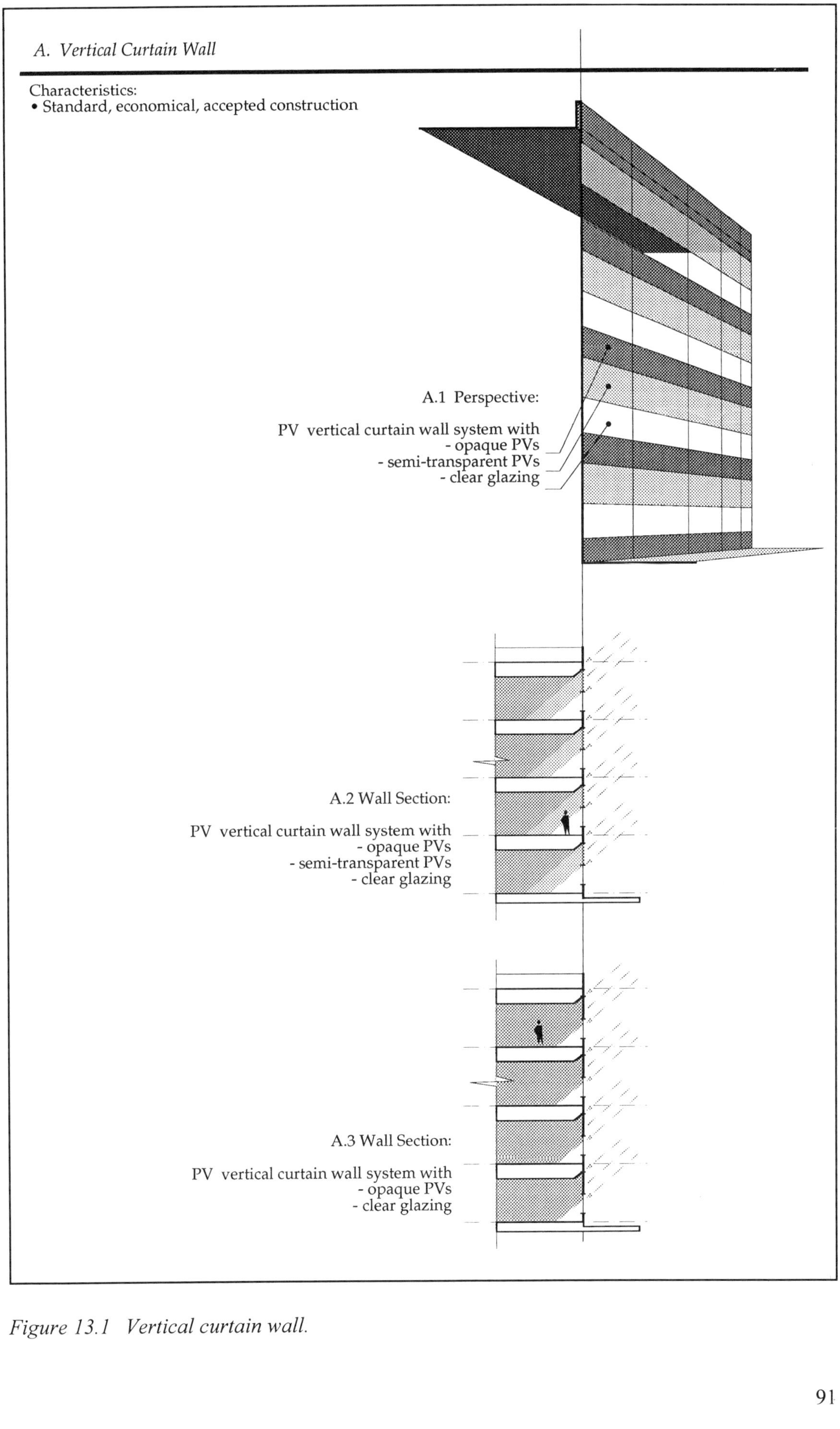


*Figure 13.1 Vertical curtain wall.*

*B. Sawtooth Vertical Curtain Wall*

Characteristics:
- Minimal additional construction cost
- Good solar performance in certain orientations
- Creates multiple "corner" windows

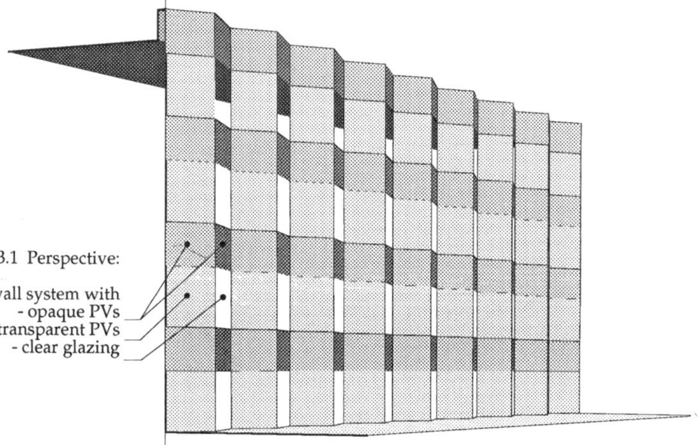


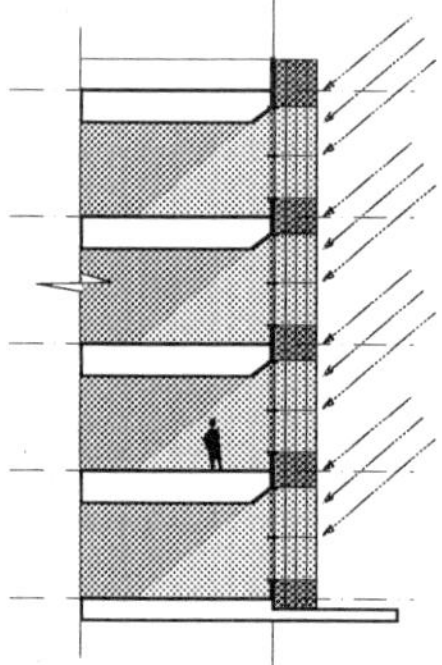


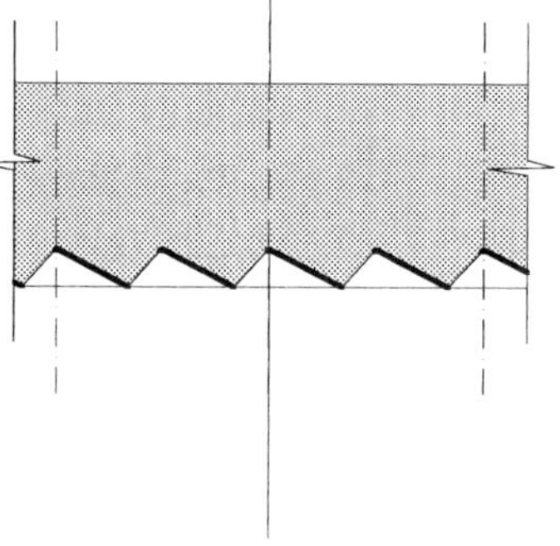


*Figure 13.2 Sawtooth vertical curtain wall.*

*C. PV Sawtooth Curtain Wall*

Characteristics:
- PVs as building skin
- Complex curtain wall construction
- Good PV efficiency
- Passive shading/daylight control
- Potential cleaning problems

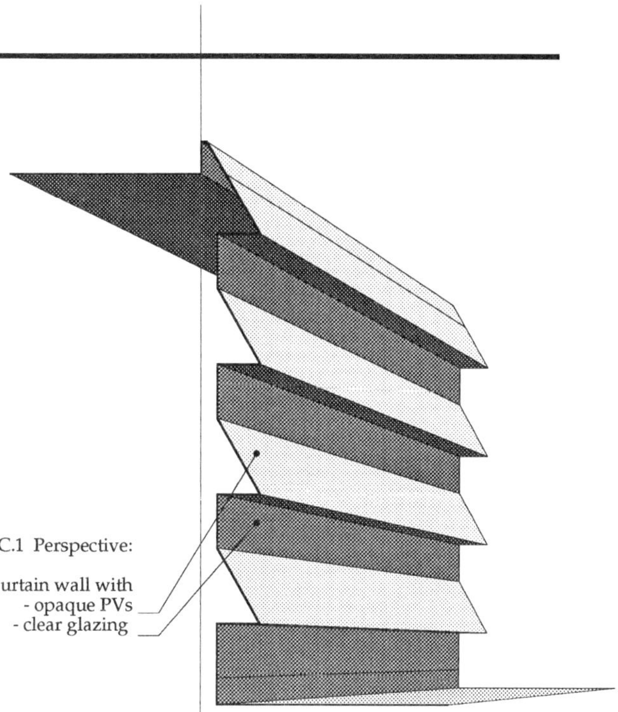


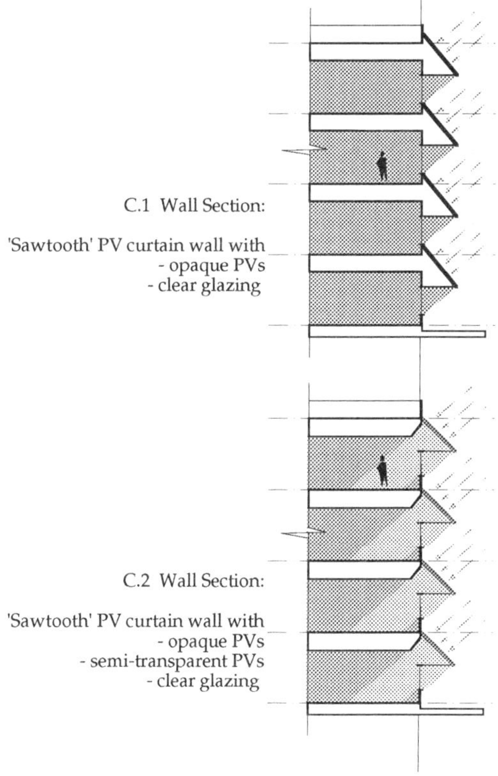


*Figure 13.3 PV sawtooth curtain wall.*

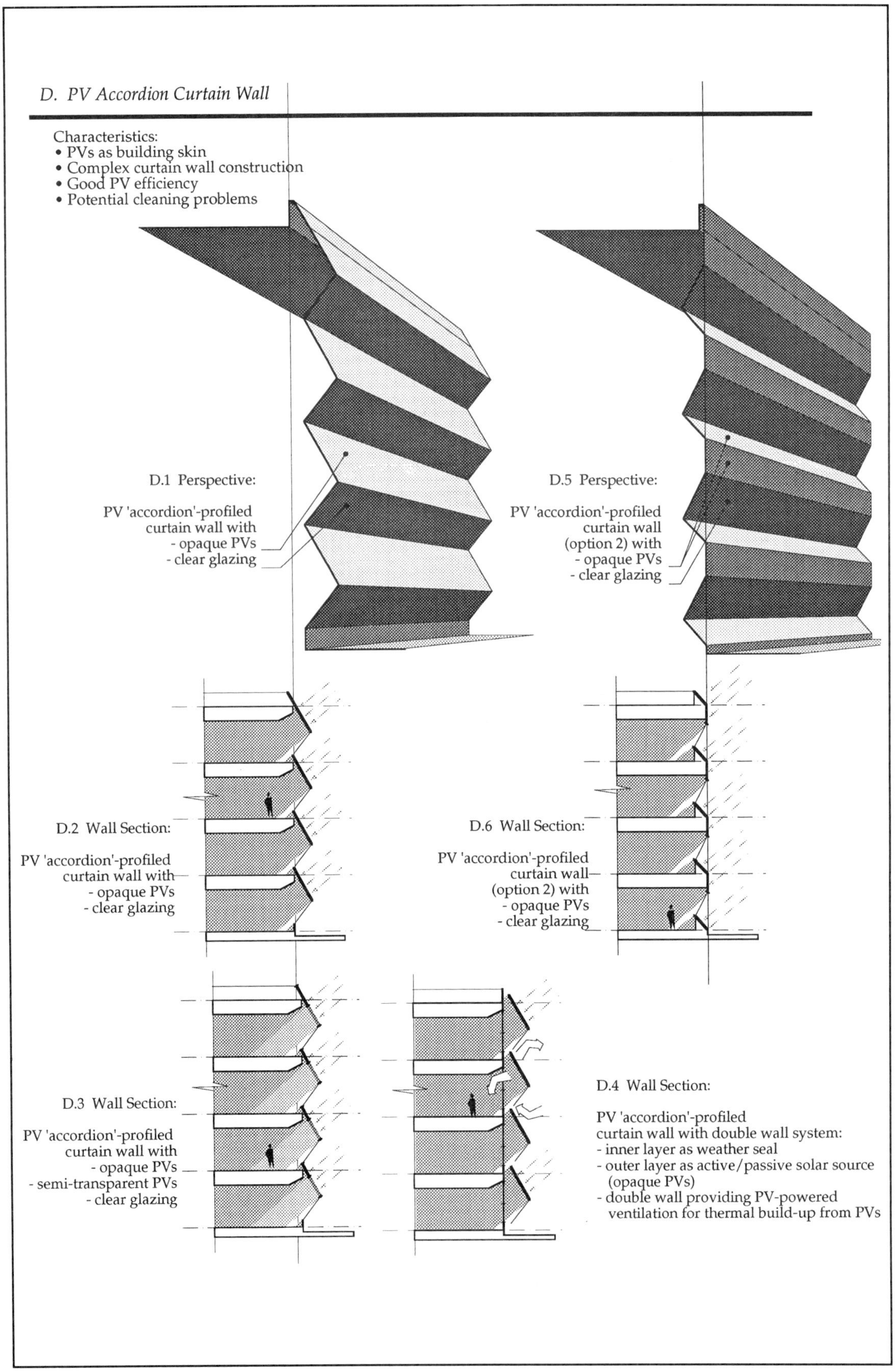


*Figure 13.4 PV accordion curtain wall.*

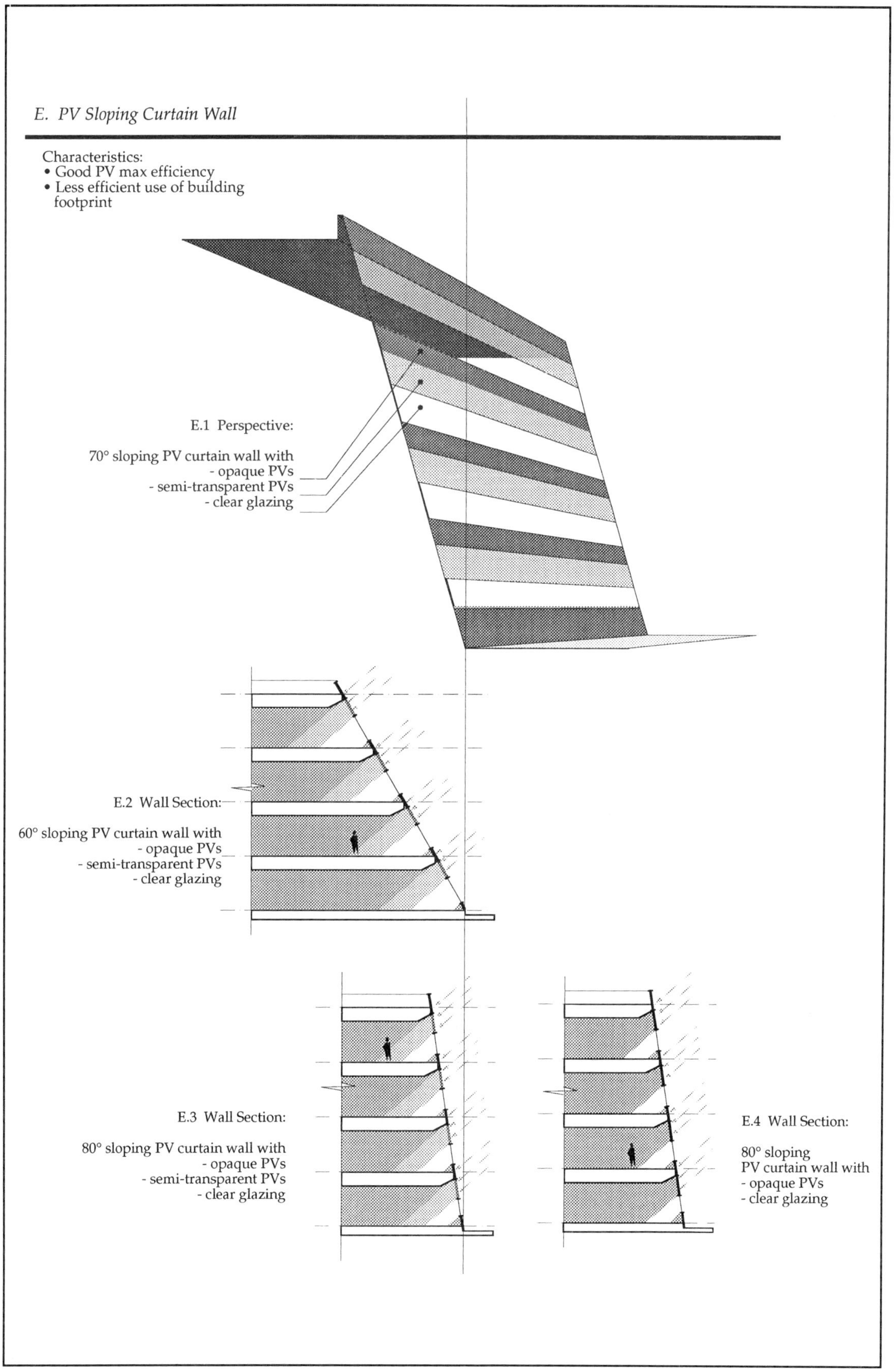


*Figure 13.5 PV sloping curtain wall.*

*F. PV Sloping/Stepped Curtain Wall*

Characteristics:
- Good PV max efficiency
- Less efficient use of building footprint
- Complex curtain wall construction

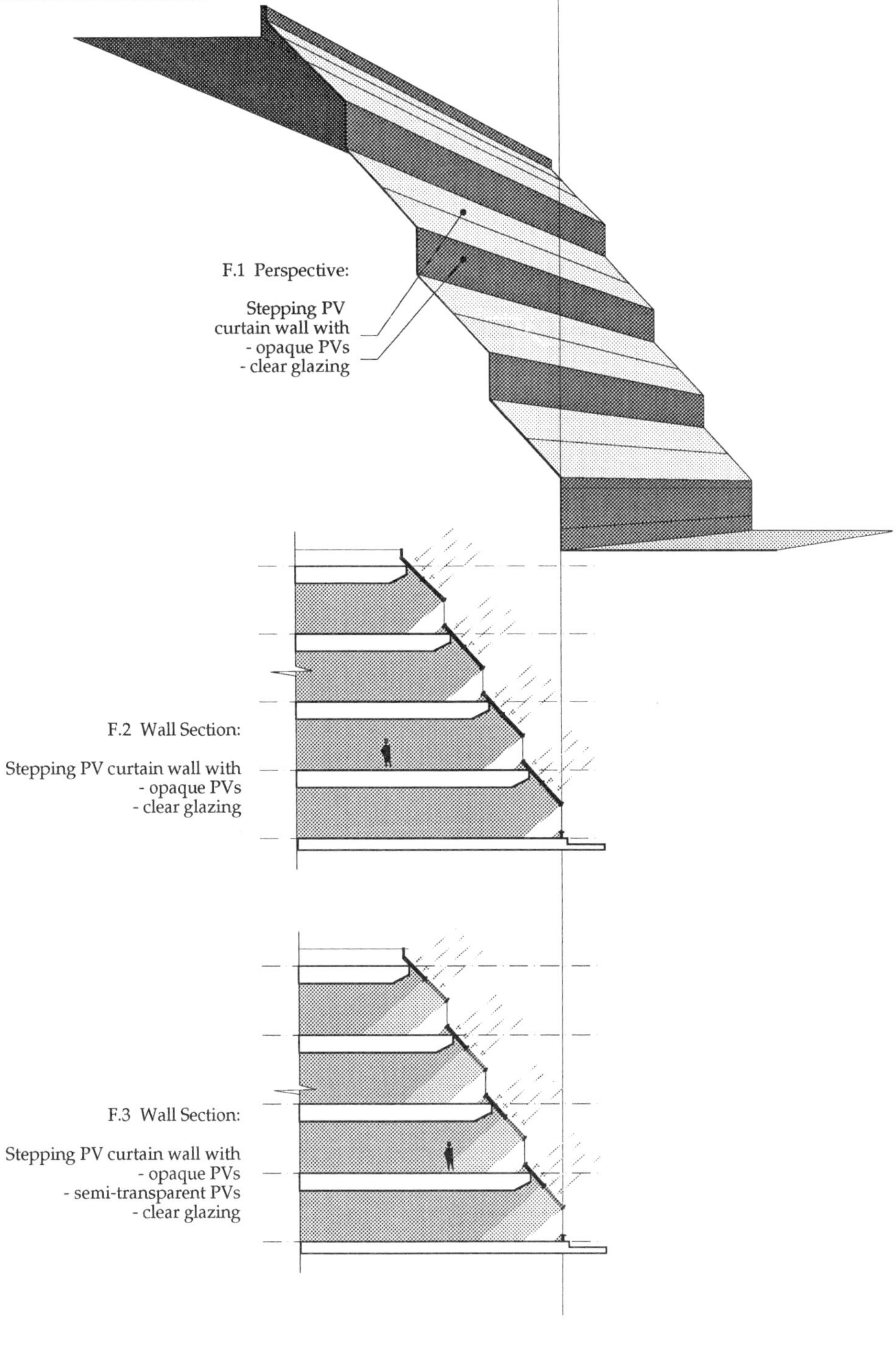


*Figure 13.6 PV sloping/stepped curtain wall.*

*G. PV Structural Glazing*

Characteristics:
- Standard, economical , accepted construction
- Difficult sealing problems for PV edges

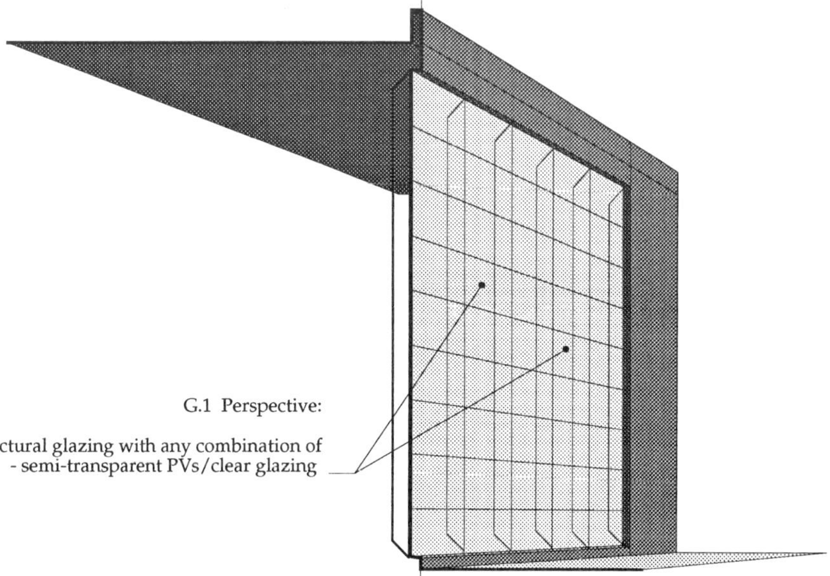

G.1 Perspective:

Vertical PV structural glazing with any combination of
- semi-transparent PVs/clear glazing

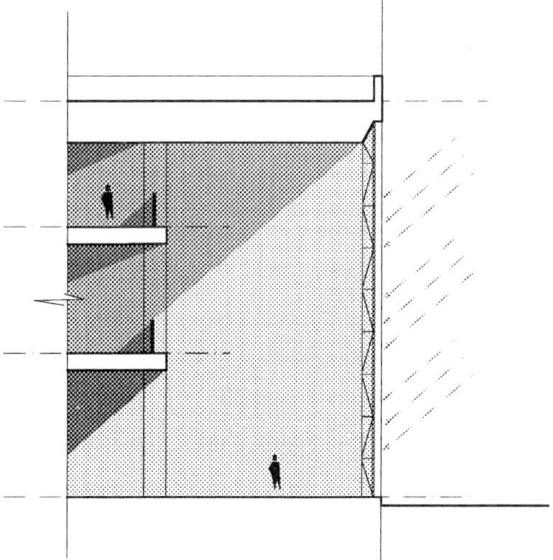

G.2 Wall Section:

Vertical PV structural glazing with any combination of
- semi-transparent PVs/clear glazing

*Figure 13.7 PV structural glazing.*

*H. PV Roof Panels*

Characteristics:

- PVs as building skin
- Combined with rooftop structural system (panelized units with insulation, fastened directly to roof structure)
- Weatherproofing and structural issues must be carefully resolved
- Snow accumulation considerations

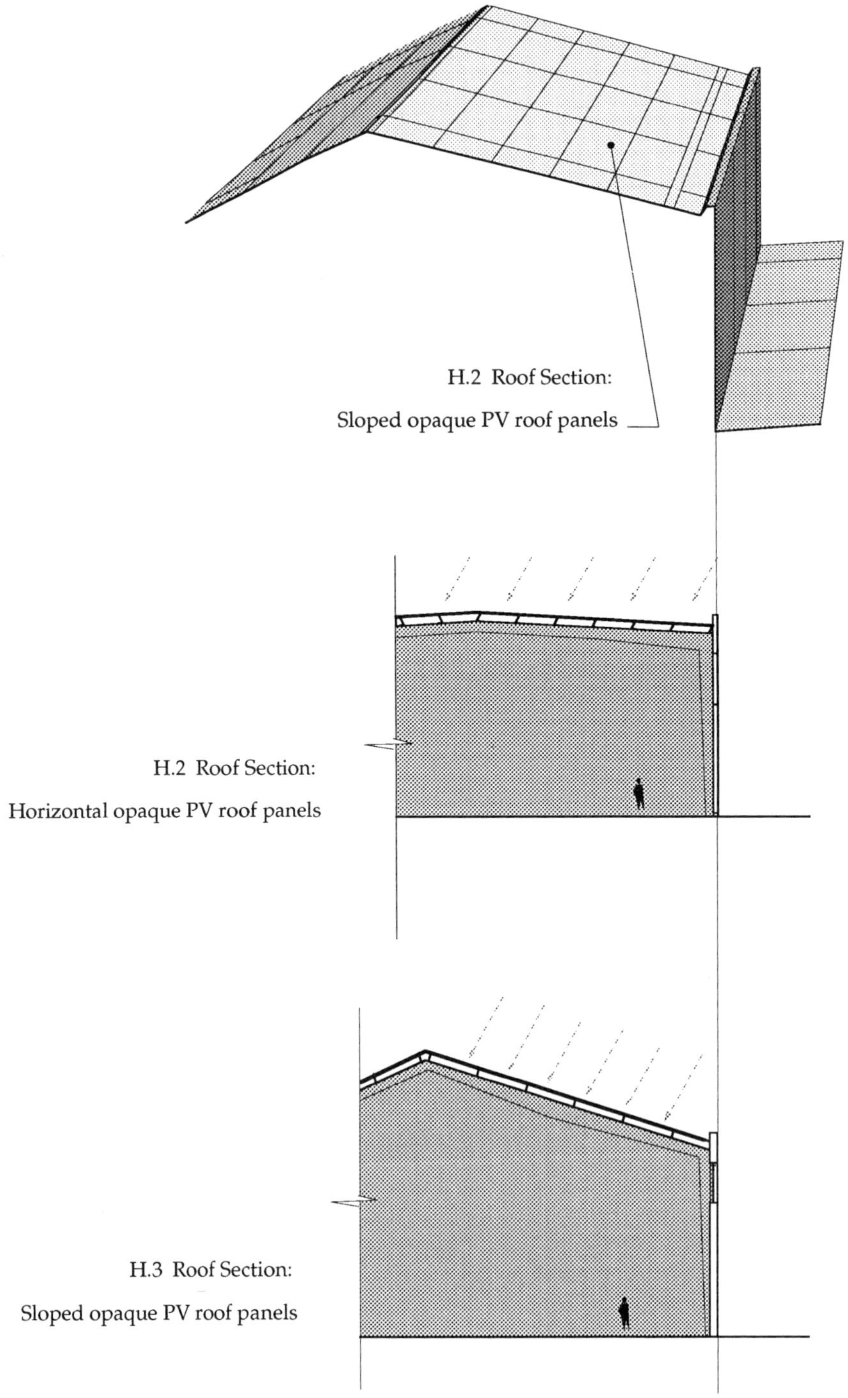


*Figure 13.8 PV roof panels.*

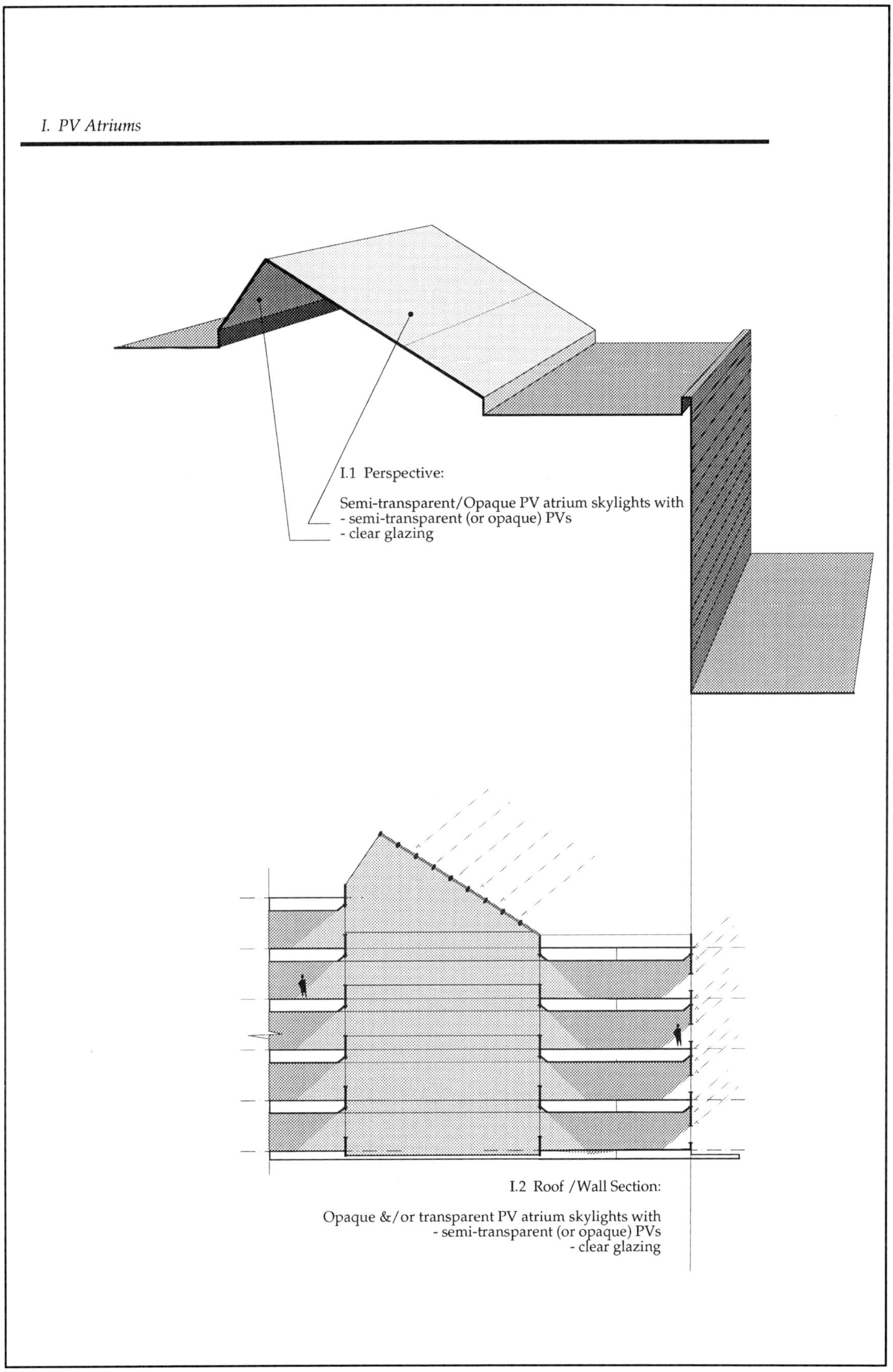


*Figure 13.9 PV atriums.*

*J. Flexible/ Metal PV Substrates*

Characteristics:
- For roofs and/or wall applications
- Good design flexibility
- Light-weight
- Possible integral weather barrier

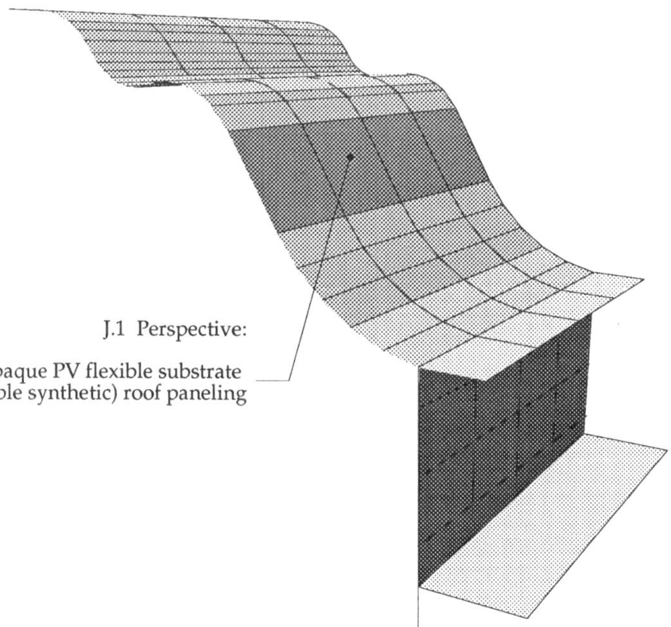

J.1 Perspective:

Opaque PV flexible substrate
(sheet metal or flexible synthetic) roof paneling

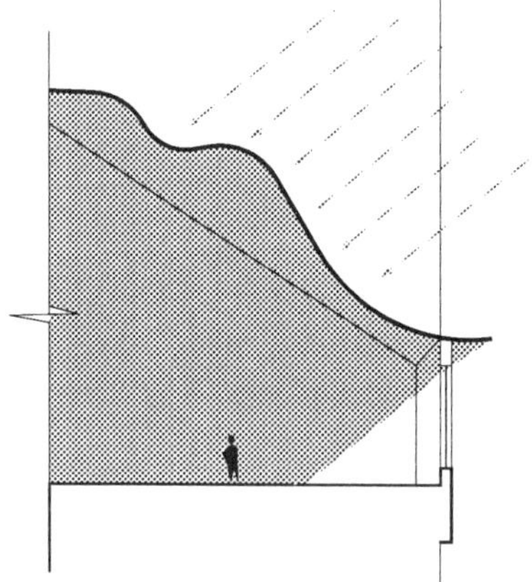

J.2 Wall Section:

Opaque PV flexible substrate
(sheet metal or flexible synthetic) roof paneling

*Figure 13.10 Flexible/metal PV substrates.*

*K. PV Skylights*

Characteristics:
- PV system as indiv. roof openings
- New construction or retrofit
- Tilted or horizontal orientation
- Numerous configurations possible
- Daylighting benefits
- Snow accumulation considerations

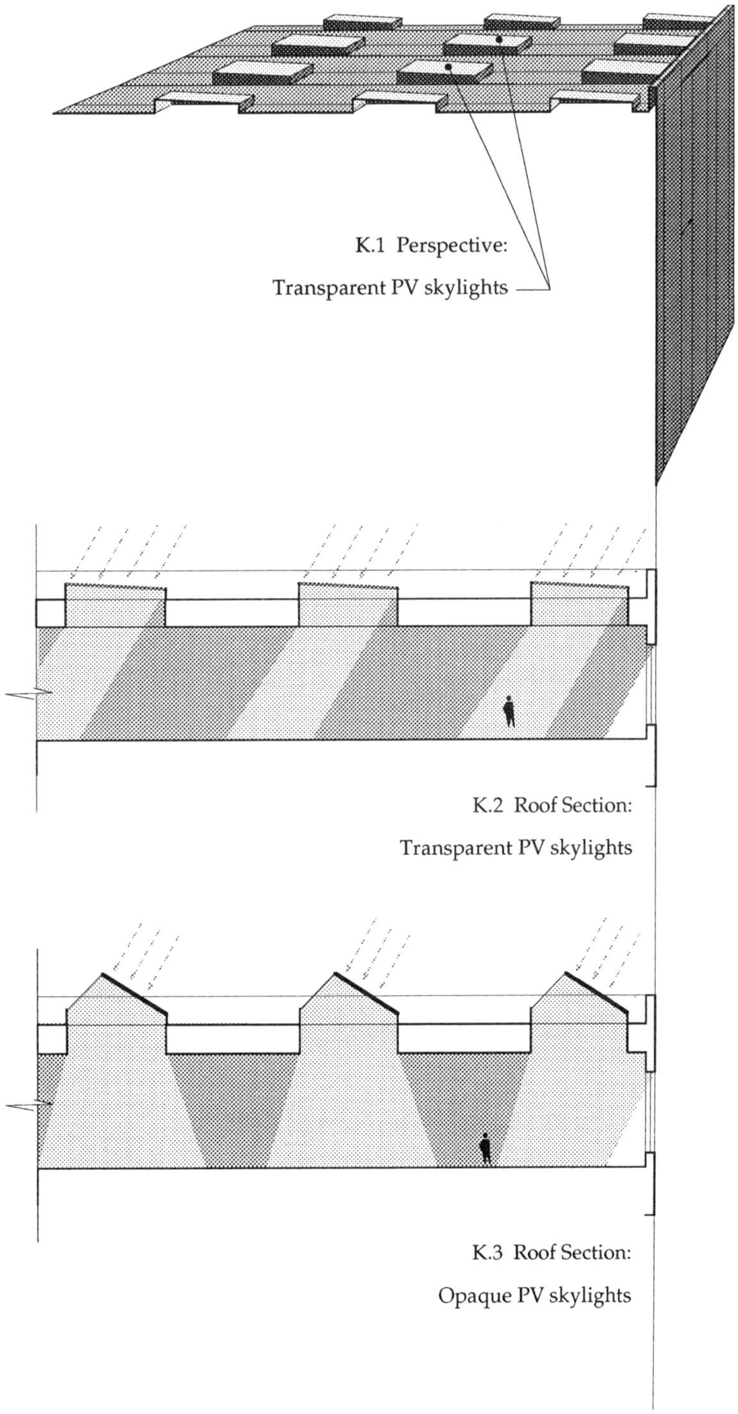


*Figure 13.11 PV skylights.*

*L. Independent PV Rooftop Array*

Characteristics:
- PV system independent of bldg skin
- conventional array configuration installed on rooftop
- Maximal efficiency
- New construction or retrofit
- Potential passive benefit from reduced heat load
- Potential structural costs
- Water proofing issues at roof/structure

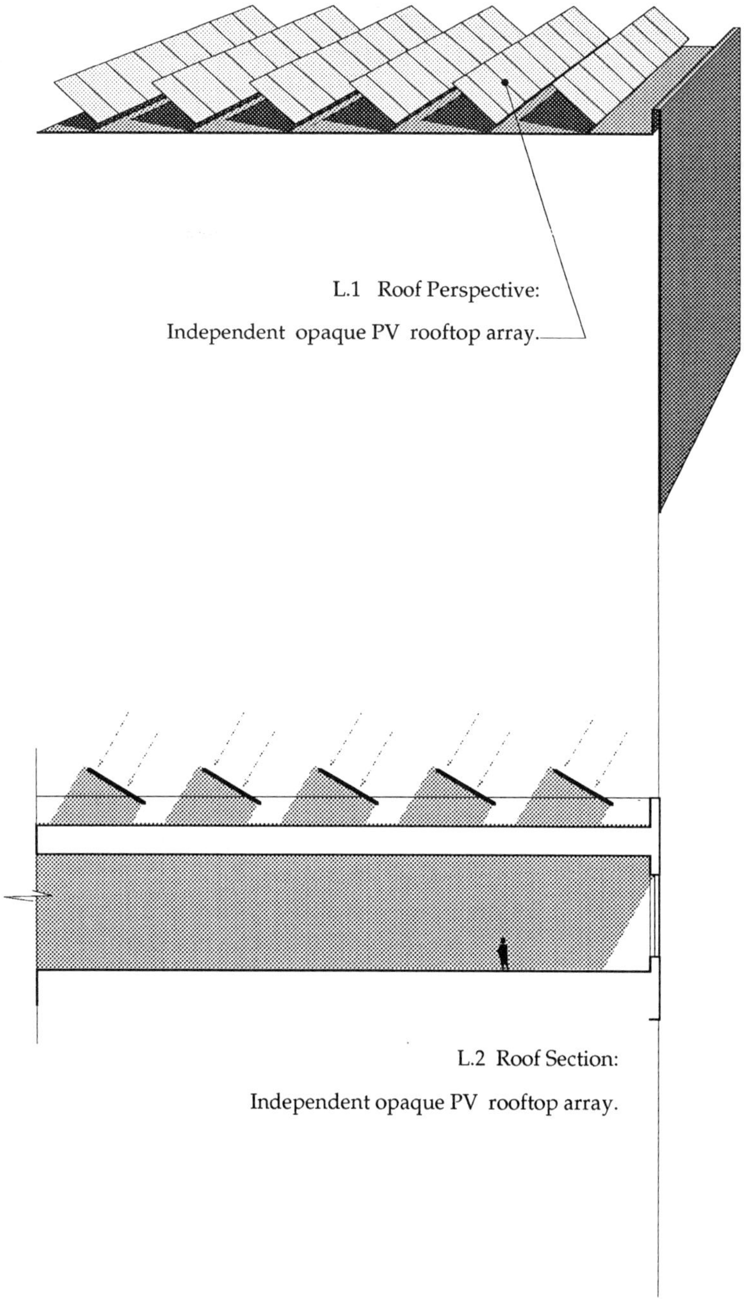


*Figure 13.12 Independent PV rooftop array.*

*M. PV Sawtooth Roof Monitors*

Characteristics:
- PV system as building skin
- Retrofit to exist. industrial buildings
- Good PV efficiency
- Good daylight benefits

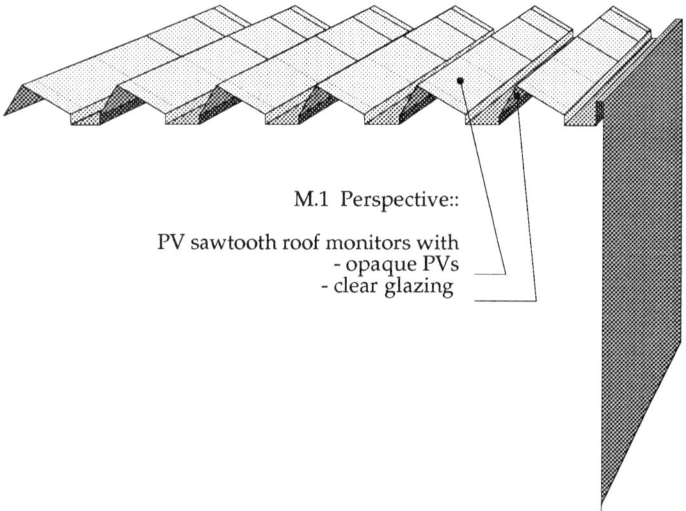

M.1 Perspective::

PV sawtooth roof monitors with
- opaque PVs
- clear glazing

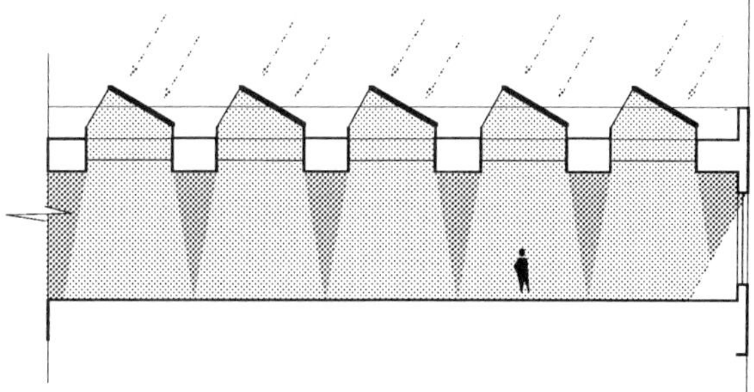

M.2 Roof Section:

PV sawtooth roof monitors with
- opaque PVs
- clear glazing

*Figure 13.13 PV sawtooth roof monitors.*

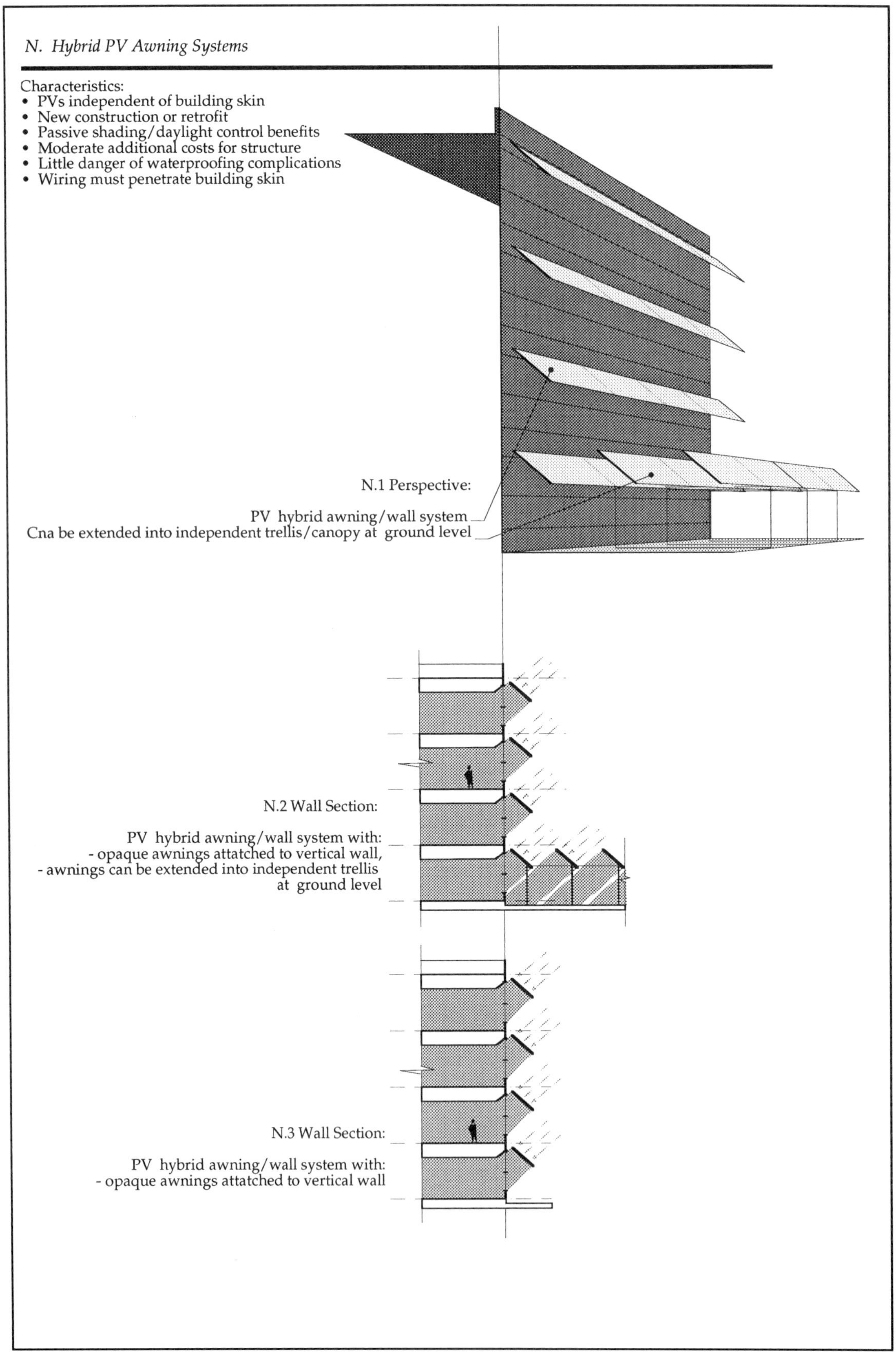


*Figure 13.14 Hybrid PV awning systems.*

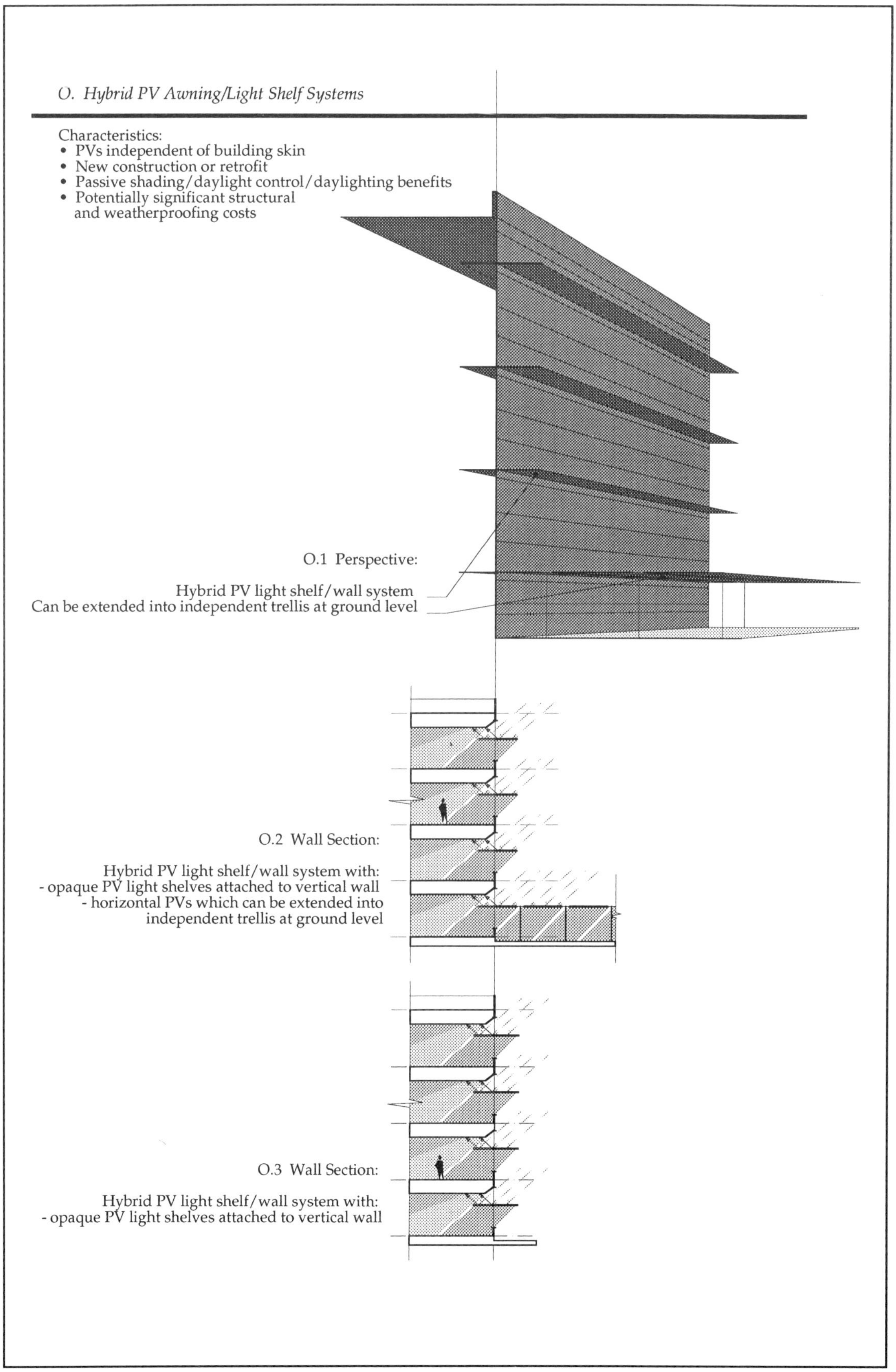


*Figure 13.15 Hybrid PV awning/light shelf systems.*

## 13.6 IEA Task 16 Architectural Ideas Competition

The following examples have been extracted from the IEA Architectural Ideas Competition 'Photovoltaics in the Built Environment'. The initiative for this competition, launched in 1993, lies with Task 16 of the IEA Solar Heating and Cooling Programme.

### 13.6.1 'SouthEastSouthSouthWest' B.J. van den Brink, The Netherlands

Environmentally-friendly building means building as little as possible, and therefore as well building as compactly as possible. A compact layout reduces the amount of needed building material. A compact building layout reduces as well the need for energy when the building is in use. Finally, compact urban planning reduces need for transportation.

The detached houses have a compact layout (they approximate a sphere), but cannot match these principles completely because of a lack of urban compactness. To compensate for this, the facades can absorb a lot of solar energy, and futuristic flaps on the roof can open during the day to collect solar energy and close during the night to insulate the building.

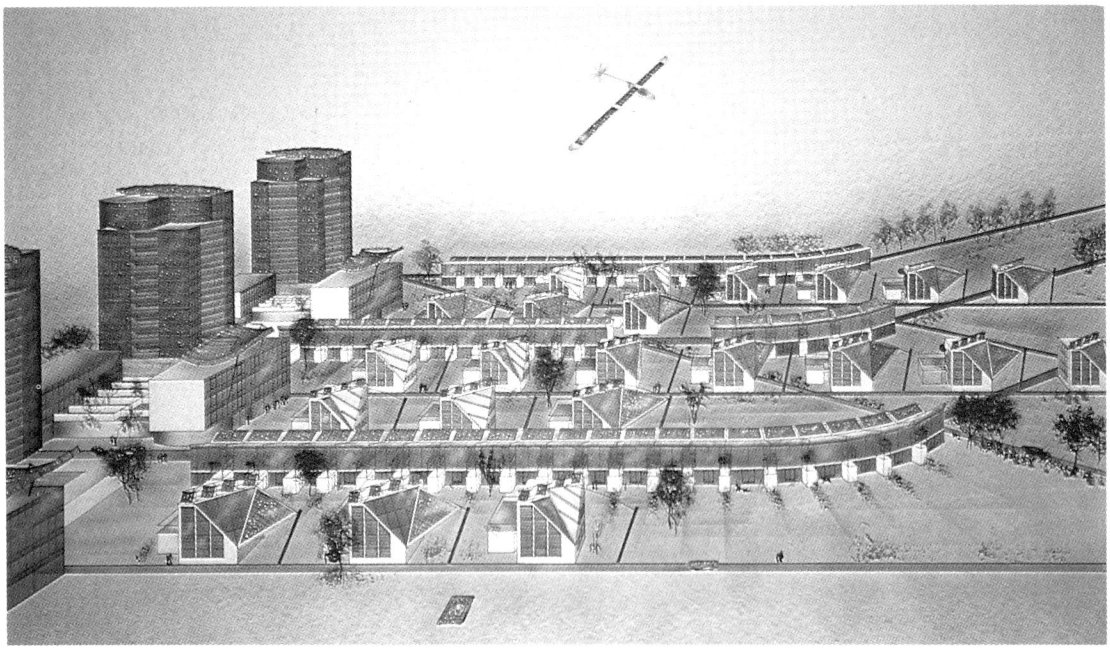

*Figure 13.16 IEA competition entry by B.J. van den Brink, The Netherlands (see colour plate 2).*

### 13.6.2 PV in single dwelling unit
### D. Mizrahi, Switzerland

The design of a single dwelling unit using PV is a very delicate question which demands a higher grade of sensibility. The perceiving of the social implications of technological development and the responsibility for our environment are indispensable for that. To avoid 'architectural pollution' it is necessary to approach PV as a self-defined design and construction element of architecture. In this way, PV becomes an obvious part of the building.

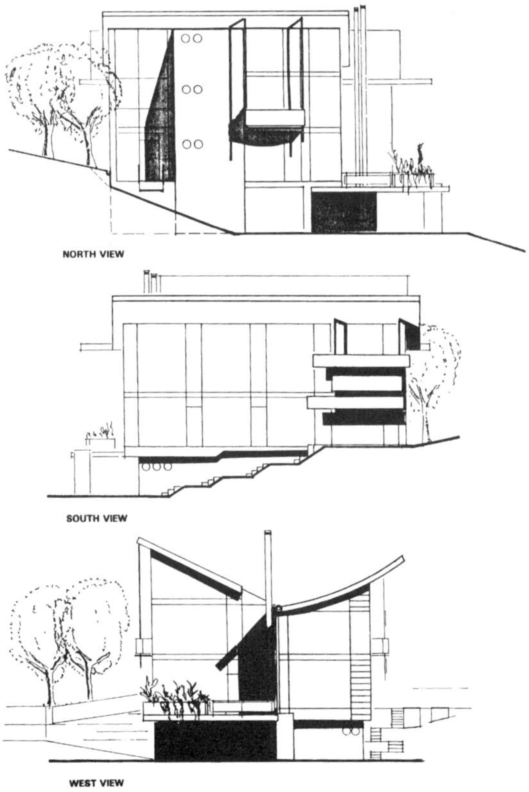


*Figure 13.17 IEA competition entry by D. Mizrahi, Switzerland.*

*Figure 13.18 IEA competition entry by D. Mizrahi, Switzerland.*

### 13.6.3 A conference and cultural centre
### G. Kiss, USA

*Figure 13.19 IEA competition entry by G. Kiss, USA.*

Large-area, thin-film photovoltaics are used throughout. Uniform appearance, large size and low cost are the principle advantages of this type of PV. In each application in this design, collateral energy benefits (daylighting, thermal) are maximized while initial costs are minimized by multifunctional construction. The conference centre roof 'sawtooth' is directly supported by box trusses tilted to support the photovoltaics while providing drainage sufficient to keep the modules clean, and providing clerestory area for full daylighting. Heat buildup (not needed for space heating in this climate) is circulated below the trusses and exhausted outside.

The hotel curtain wall is internally ventilated from floor to floor by convention dampers at the head and sill of each window. Hot air from the wall cavity is directed in or out according to demand. This avoids the costs of a full double wall system while providing the flexibility of operable windows.

*Figure 13.20 IEA competition entry by G. Kiss, USA.*

### 13.6.4 An Energy-Conscious Office Building in Ankara
### C. Elmas, USA

In Turkey, integration of photovoltaics in buildings has not been performed yet. In order to make this technology acceptable, this project uses the PV modules as sunshades that are reminiscent of traditional *mashrabiyas* (wood lattice screens). PV modules are used with different configurations, transparencies and patterns at each facade in order to respond to the sun, views from the offices, character and scale of the particular facade.

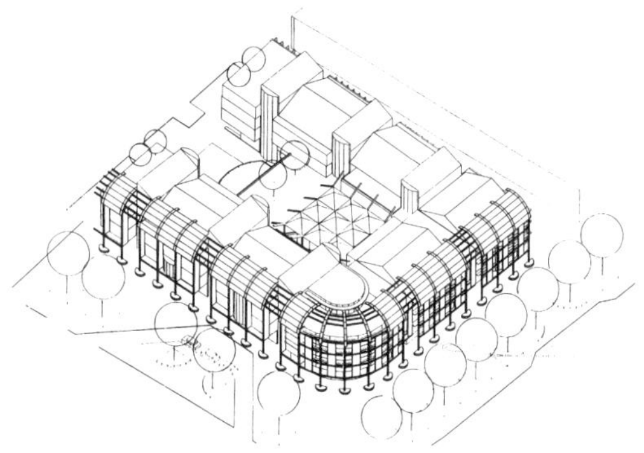

*Figure 13.21 IEA competition entry by C.Elmas, USA.*

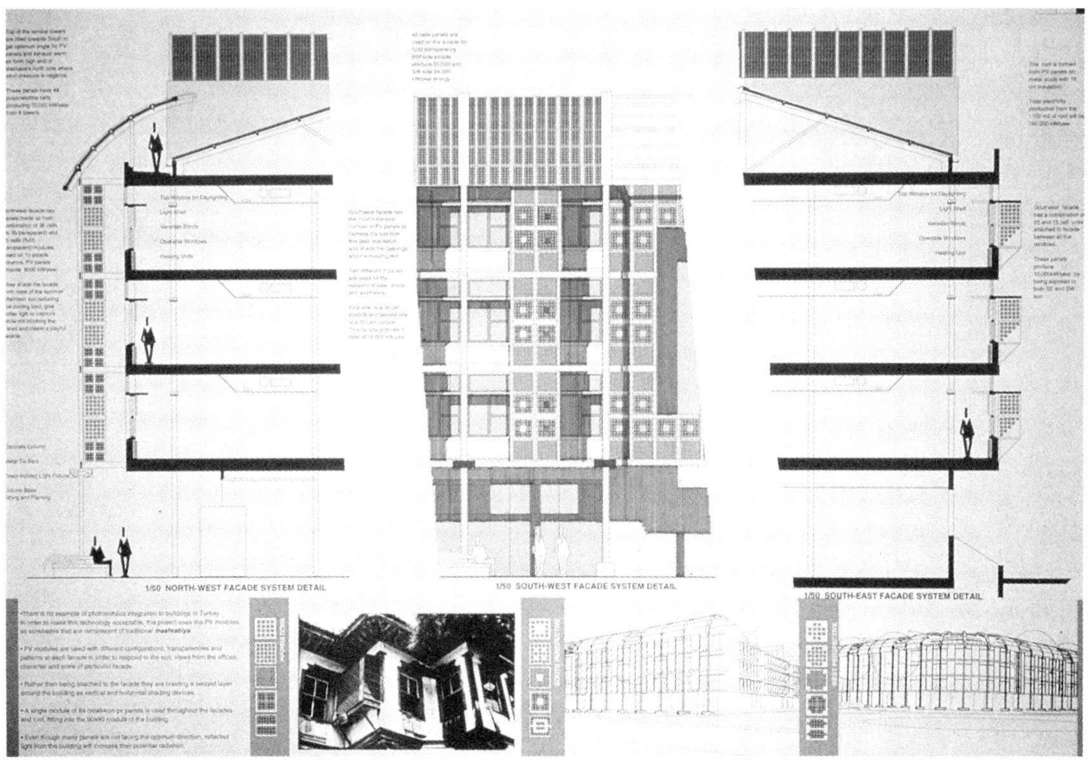


*Figure 13.22 IEA competition entry by C. Elmas, USA.*

## 13.7 German Architectural Ideas Competition 1994

The architectural ideas competition 'Photovoltaics in Buildings' was initiated to encourage architects in cooperation with engineers to tackle the technically and aesthetically challenging problem of building integration of photovoltaic systems. It was supported by the Federal Ministry of Education, Science, Research and Technology under a grant to the Fraunhofer Institute for Solar Energy Systems (FhG-ISE) in Freiburg, Germany, and organized by the Institute for Industrialization of Building Construction (Institut IB) in Hanover, Germany.

Using a building that was in planning and/or under construction anyway, possibilities for the integration of photovoltaic systems with high architectural quality were to be shown. The conditions for and the consequences of the integration were to attract special attention to the building. The buildings could be of any size and location within Germany, e.g. office and administration buildings, traffic facilities, sports facilities, convention halls, industrial buildings, schools or residential buildings.

The jury met in September 1994 to evaluate the 30 competition entries. The entries were judged in three rounds according to the following criteria:

- the overall architectonic concept,
- the overall energy concept,
- the integration of the photovoltaic elements,
- the transferability of the solution,
- the expense and the gain from the photovoltaic systems used.

Seven entries were awarded and mentioned:

### 13.7.1 Zentrum für Kunst und Medientechnologie, Karlsruhe Schweger + Partner, Hamburg

An industrial area from the turn of the century will be modernized and prepared for the Center for Arts and Media Technology. The photovoltaic generators will be integrated in the transparent roof of the patios. This gives a multiple effect:

- reduction of energy entry and air-conditioning power by about 90 kW,
- a photovoltaic generator with a nominal power of about 100 kW,
- direct feeding of the electrical power into the overhead line of the Karlsruhe trolley cars (no batteries or current inverters).

*Figure 13.23 Visible photovoltaic panels above the entry hall (see colour plate 5).*

*Figure 13.24 Cube of music studio outside the existing building.*

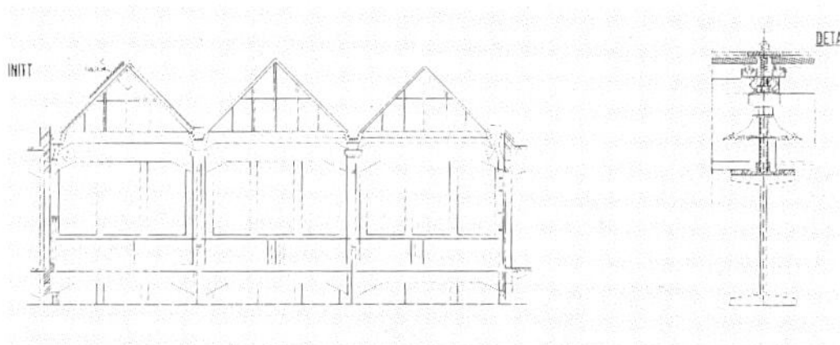

*Figure 13.25 PV array on top of the music studio.*

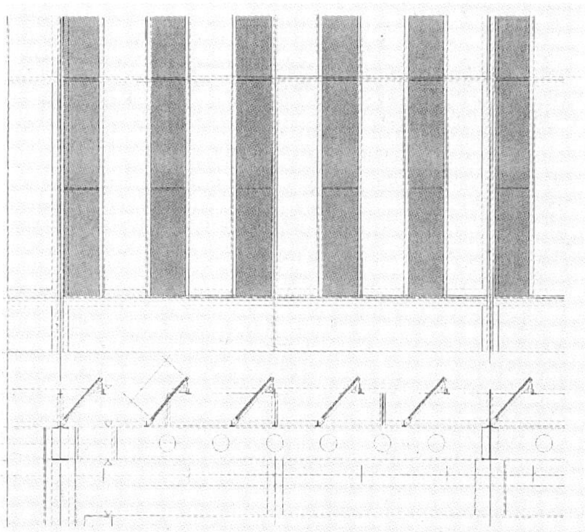

*Figure 13.26 Section of the glass roof.*

### 13.7.2 Exhibition building 'Haus des Waldes', Stuttgart
### Dipl.-Ing. Michael Jockers, Stuttgart

The pavilion contains an exhibition hall and supporting facilities. It is located in the Stuttgart city forest close to the television tower. The photovoltaic generator (14.5 kW) is used in addition to other energy saving measures (passive heat gains, daylighting, high insulation of opaque components, use of wood as load-bearing material, solar collectors for hot water supply).

*Figure 13.27 Model view of 'Haus des Waldes'.*

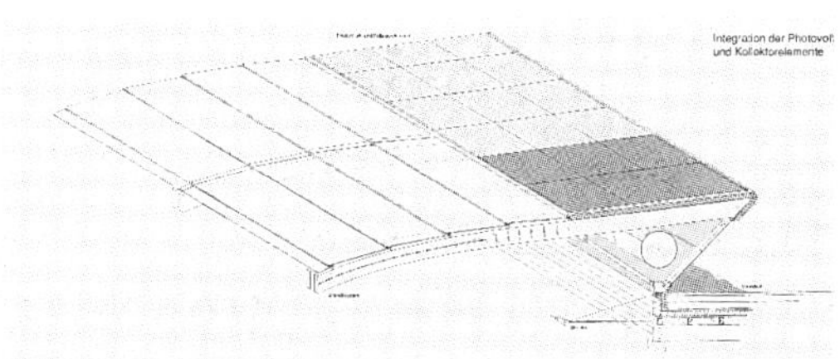

*Figure 13.28 Detail of roof integration ('Haus des Waldes').*

### 13.7.3 Housing project with nine apartments in Aalen Kerler-Amesöder-Braun, Fellbach

The terrace-shaped apartment building shows an interesting way of integrating the photovoltaic generator in combination with windows and loggias. The system is dimensioned to cover approximately two thirds of the yearly electrical energy consumption. Even during the summer months, almost all of the energy generated is thus consumed within the building. Excess energy is fed into the mains.

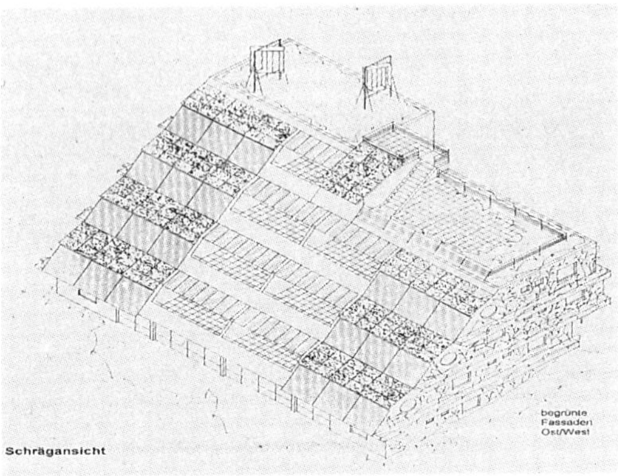


*Figure 13.29 Section terrace.*

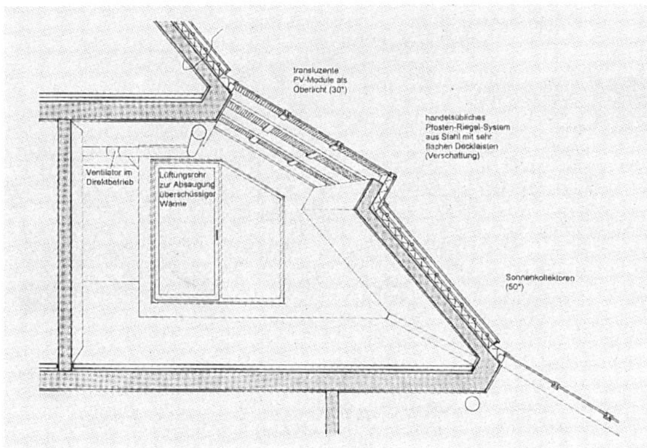


*Figure 13.31 Section room.*

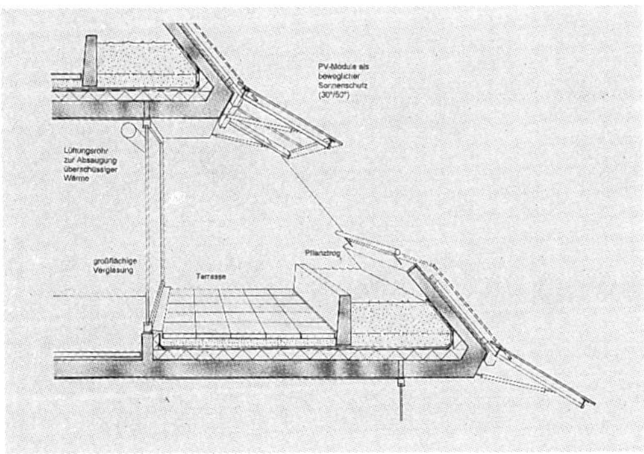


*Figure 13.30 Bird's eye view.*

### 13.7.4 Administration building for Nürnberger Versicherung, Nürnberg
**P. Dirschinger, F. Bifang, Ammerndorf**

This is a large administration building (about 4500 employees) with photovoltaic generators integrated into different components (among others noise-absorbing walls, shading devices and an advertising sign). The total nominal electrical power is 366 kW. Other measures related to the energy consumption are transparent insulation, thermal collectors, daylighting, cold buffering, heat recovery and optimal insulation of all outer walls.

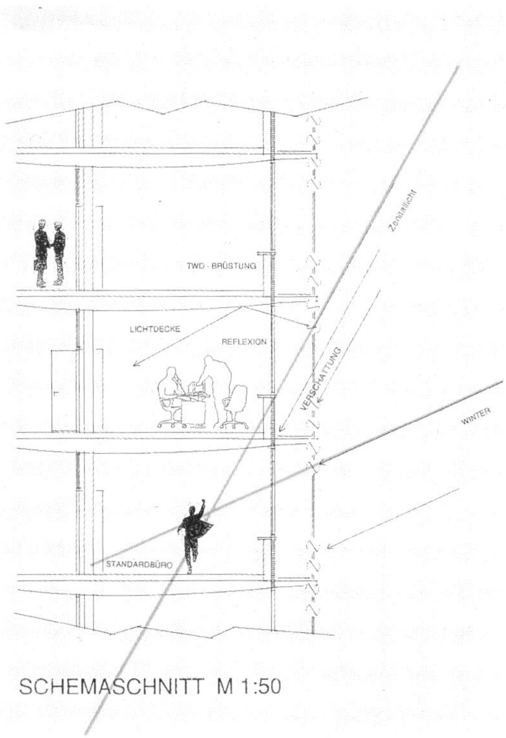


*Figure 13.34 Section with shading PV elements.*

*Figure 13.32 Model view.*

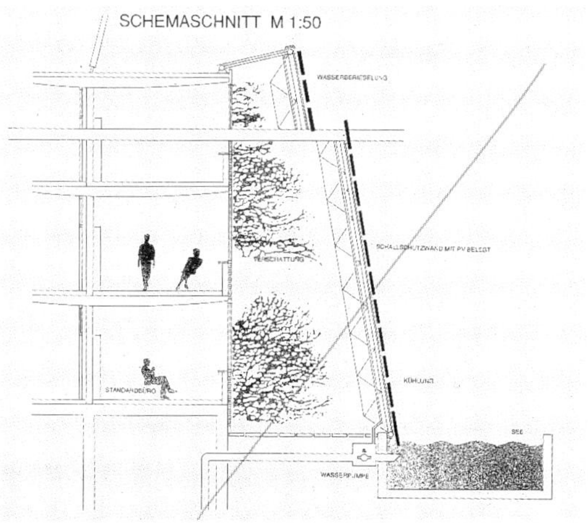


*Figure 13.33 Section of sound protection wall with PV array.*

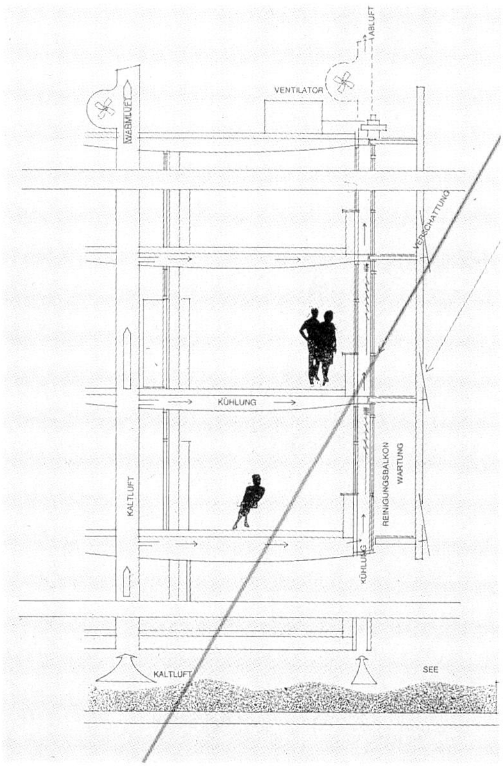


*Figure 13.35 Section of ventilated facade.*

### 13.7.5 Office building Luxemburger Straße, Hürth
### Elkin + Hövels, Cologne

This building is planned for a private investor on the outskirts of Cologne on a corner site with good traffic connections. Several energy-relevant measures are planned:

- high-insulating glass facade,
- ventilation with heat recovery,
- integrated photovoltaic generators ('saddleback roof', 'great sun shovel', 'little sun shovel').

*Figure 13.36 View from east.*

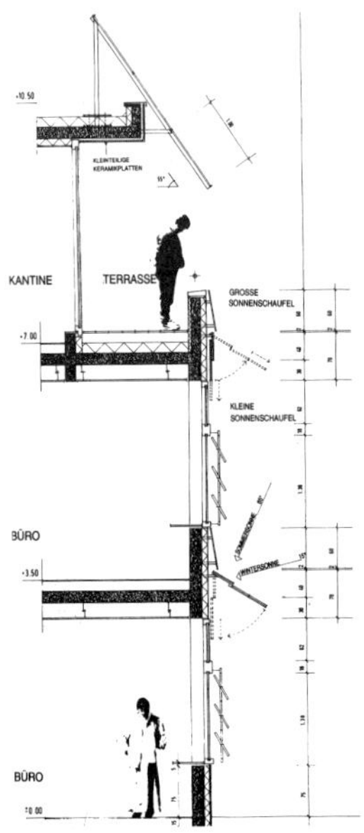


*Figure 13.37 Facade section.*

### 13.7.6 Ecological housing estate, Hamm-Heessen
### H.F. Bültmann, Bielefeld

This project, which will be built in two phases, will provide about 100 housing units of different types. The authors plan to approximately cover the yearly electrical energy consumption of the estate with a photovoltaic system of about 196 kW. Excess electrical energy will be fed into the mains.

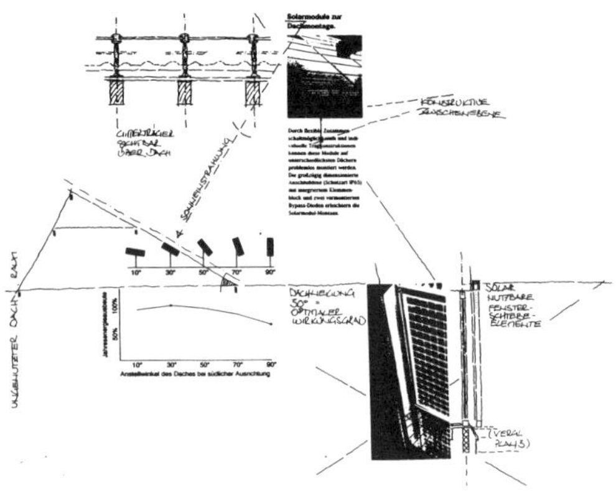

*Figure 13.38 Integration details.*

*Figure 13.39 Western and eastern facades.*

### 13.7.7 Radio relay station Tabarz I, Botterode/Thüringen Streckebach Architekten, Kassel

In this pilot project, a photovoltaic generator and batteries are used as backup to increase the reliability of the power supply. It is investigated as a possible way to replace diesel generators in environmentally sensitive areas. An architectonically interesting way of mounting the photovoltaic generator to the existing tower was found.

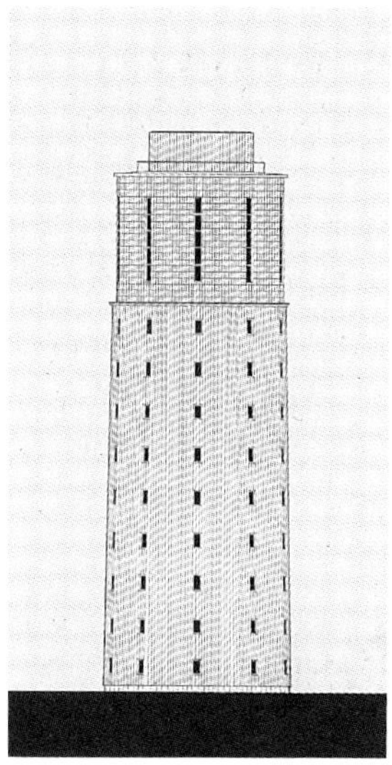

*Figure 13.40 Existing tower.*

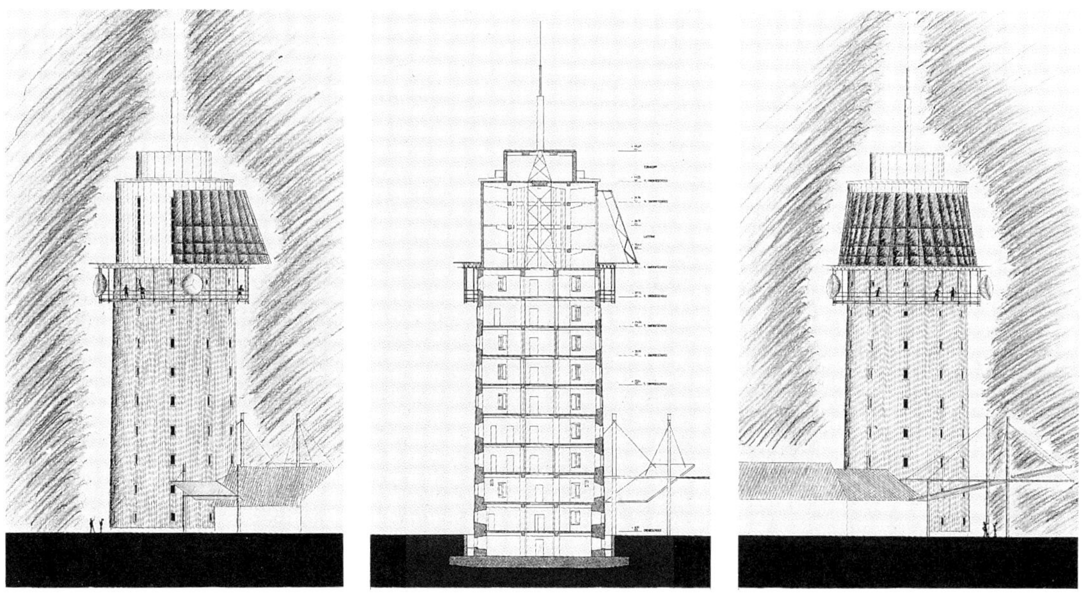

*Figure 13.41 Tower with PV array.*

## Chapter 14

# Integration Techniques and Examples

### 14.1 Introduction

The following selection of examples aims at presenting basically two key issues. One is the demonstration of a significant number of realized examples: facades, roofs and others show clearly that the architectural and technical integration of PV in buildings is feasible. And as a second message, the descriptions show how the systems are mounted. Each mounting method is explained by a set of drawings. In addition to these example presentations, some general recommendations for module integration are given as follows.

**Use of non-corrosive materials:** On PV facades there are always small leakage currents around. In order to avoid corrosion caused by the leaking DC currents, it is essential to use non-corrosive construction materials. The joints and holding points demand special concern.

**Exchange of modules:** The exchange of single modules should be possible without taking away a large part of the facade or roof.

**Flush surfaces:** To avoid shading, dust accumulation and to ease clearing by rain, protruding elements over the solar modules should be avoided. Fixation points for an unframed module may only little jut out the module surface.

**Easy cabling:** Before the solar roof or facade elements are placed, the electrical connections have to be prepared. The use of reliable plug connectors makes a quick and easy installation possible. During mounting of the modules no electricians should be required.

**Minimum module handling on site:** As most of the current solar modules consist of a glass surface, they require careful handling during transportation. Once installed, the risk of damage is very low. This circumstance leads to the recommendation that module handling on the construction site should be reduced as far as possible. The module should go from the transportation container directly to a position along the roof or facade.

**Module matching and string layout:** Depending on the electrical layout and the module characteristics, they have to be assigned to a certain position in order to form groups or strings of modules with similar electrical characteristics. This can be done in the factory before shipping. The packing should then be accordingly in order to ease module handling on site.

**Ventilation:** As long as modules based on crystalline cells are used, it is an advantage when the modules' rear side is ventilated as with curtain walls. With amorphous modules the difference is negligible because the efficiency depends only a little on temperature.

## 14.2 Stadtwerke Aachen, Aachen, Germany

*Figure 14.2.1 General view (see also colour plate 4).*

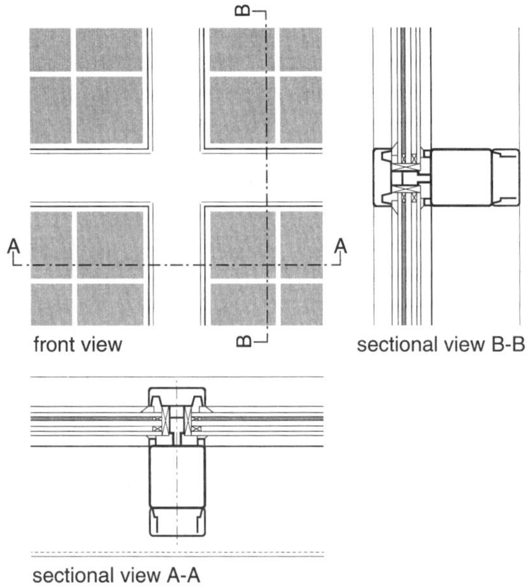


*Figure 14.2.2 Detail of mounting structure.*

**Key data**

| | |
|---|---|
| Nominal power: | 4.2 kW |
| Gross area: | 87 $m^2$ |
| Module manufacturer: | Flachglas Solartechnik |
| Inverter: | 2 PVWR 1800 |
| Start of operation: | 1991 |
| Owner: | Stadtwerke Aachen AG |

**Short description**

The south-facing glass facade of the 20-year-old administration building of the Stadtwerke Aachen (Utility of Aachen) was replaced by a solar facade during the renovation of the heating services. Light-diffusing modules developed especially for this south-south east frontage direct daylight into the staircase behind. The chessboard type combination of glass elements and the modules with dark-blue crystalline silicon cells between the compound glazing offers surprising patterns outside as well as inside.

This example was one of the first so-called multipurpose applications of PV modules. The PV modules act as semitransparent facade, as wall element including thermal insulation and as power production device. The wiring is integrated in the structure holding the modules.

*Figure 14.2.3 View of the facade from the inside.*

## 14.3 ISET, Kassel, Germany

*Figure 14.3.1 General view.*

*Figure 14.3.2 Detail view of the installation behind the facade.*

**Key data**

| | |
|---|---|
| Nominal power: | about 5 kW |
| Module manufacturer: | ASE GmbH |
| Inverter: | 3 x NEG 1600 |
| | 5 x SLD 224 |
| | 3 x Weeber MIC |
| | 1 x SMA MOC |
| Start of operation: | 1.7.1993 |
| Owner: | ISET e.V., Kassel |

**Short description**

The main idea of this project is the realization of an experimental photovoltaic (PV) facade at an existing building at the University of Kassel. The facade consists of a PV array of 50 $m^2$ (75% south, 15% west and 10% east orientation) and is used for testing different components (modules and inverters) and concepts (mounting and electrical installation) in practice. Since July 1, 1993, five different PV generators have been in operation; their conditions are measured and recorded.

The project is funded by the Ministry of Environment of Hessia. The industrial project partners are ASE GmbH (DASA and NUKEM) and SCHÜCO International.

**Detail of mounting structure**

In this project existing industrial components are used for building up the PV facade. So a facade system was modified to integrate the PV modules. Three different sizes of modules were inserted: 1780 x 1180 mm, 1180 x 1180 mm and 580 x 1180 mm. The modules contain 160, 80 or 40 photovoltaic cells in order to be electrically connectable.

**Main research aspects**

The PV facade is exclusively designed for the following research aspects which enables answers to the questions and concerns of module manufacturers, facade builders, architects and energy industries. These are:

- edging quality of frameless PV modules,
- defect recognition in the PV facade system,
- measuring technology, e.g. reference measuring,
- safety of persons and the system,
- energy processing, test of different types
- meteorology - recording of weather data
- installation - mechanical and electrical concepts,
- shadowing and the influence of hot-spot-effect,
- mechanical tests - behaviour under extreme wind conditions,
- electrical tests - efficiency and insulation resistance.

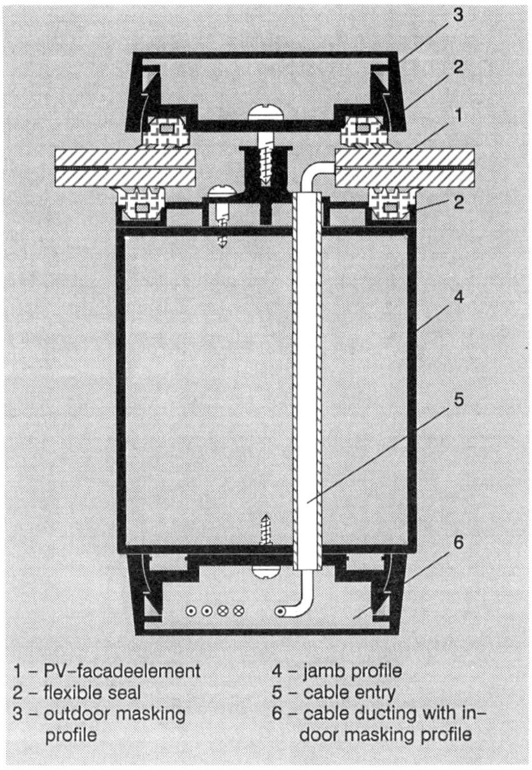


*Figure 14.3.3 Detail view of a mechanical construction (new concept).*

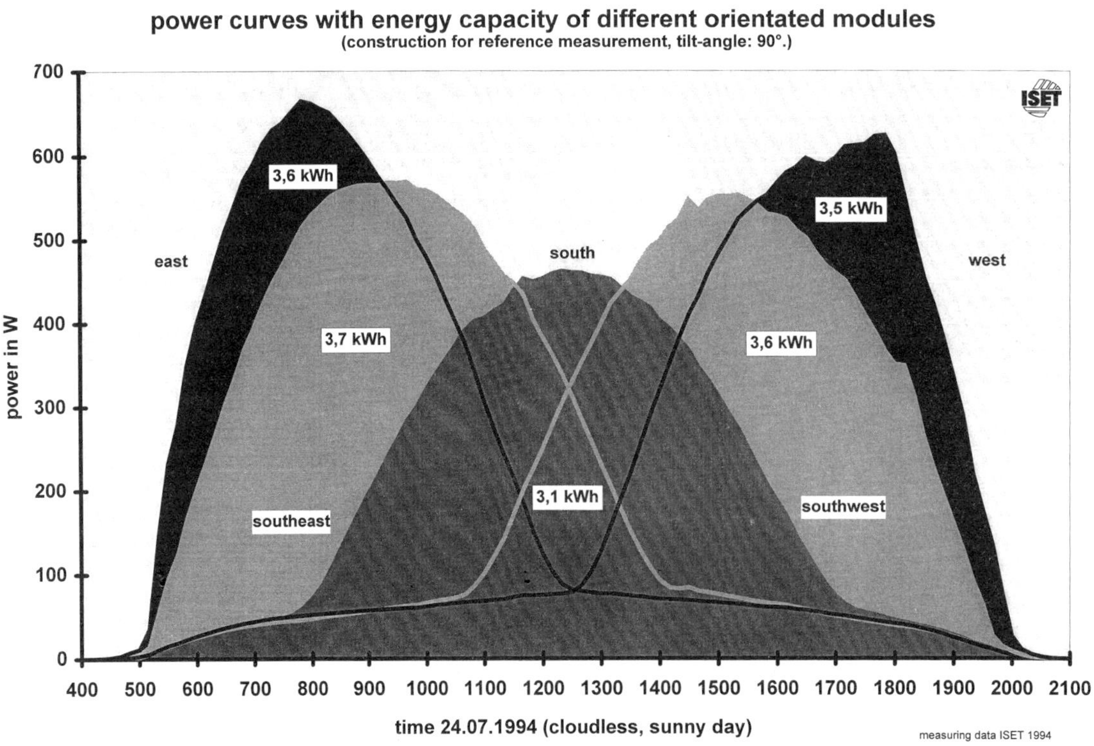


*Figure 14.3.4 Energy input of reference PV module (referring to 1 $kW_p$; location: Kassel, Germany).*

## 14.4 Ökotec 3, office building, Berlin

*Figure 14.4.1 General view (see also colour plate 6).*

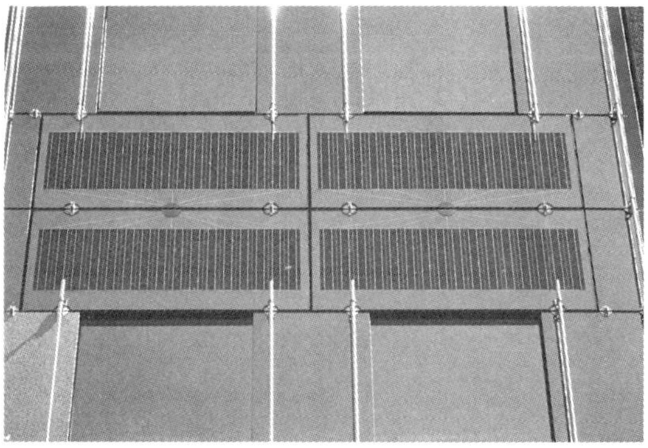

*Figure 14.4.2 Detail of PV element (see also colour plate 7).*

**Key data**

| | |
|---|---|
| Nominal power: | 4.2 kW |
| Gross area: | 64 m$^2$ |
| Module manufacturer: | Flachglas Solartechnik GmbH |
| Inverter: | 2 x PV-WR 1800 S |
| Start of operation: | 1994 |
| Owner: | Triangel Grundstücksverwaltungsgesellschaft |

**Short description**

The goal to be achieved was to integrate a solar facade of the most progressive technical state into a new industrial building with the latest architecture. Special attention should be paid to achieving high efficiency of the installation and to integrate the PV generator properly into the architecture.

The industrial building created by the architects Aicher, Jatzlau von Lennep and Schuler in Berlin-Kreuzberg, with a floor surface of 8100 m$^2$ was equipped with a grid-connected photovoltaic generator integrated in its south front. The 44 solar facade elements are integrated as cold facade elements into the framework system of a reflecting whole glass facade with interwork side ventilation.

The new 'SJ' facade system used for the first time for a solar facade project stands out for its easy mounting system for glass panels. The solar facade modules are designed as a blue glistening, partly reflecting, double pane laminated glazing unit to harmonize best with the remaining reflecting 'structural glazing' facade. The electrical connection is done by a standardized plug socket system.

With regard to the module configuration and production, the connection system - held as a standard for future projects - as well as the aesthetic successful architectonic integration, this project sets an innovative example.

The facade system is particularly suitable for the integration of PV facade modules, as it holds frameless facade panes of any kind by a minimum of four fastening points through patented, star shaped stainless steel parts with elastic EPDM coating and is thus able to carry the frameless OPTISOL module assembly. Electrical connection: The collecting line is attached to the interior pane covered with solar control reflective insolating coating. The bypass cables coming from the panel board through standard sockets are connected with the modules. These modules are interconnected with a socket/plug system, the junction box being stuck to the glass rear side of the module. The whole connection system makes the installation work easier and, if necessary, enables easy replacement of each module.

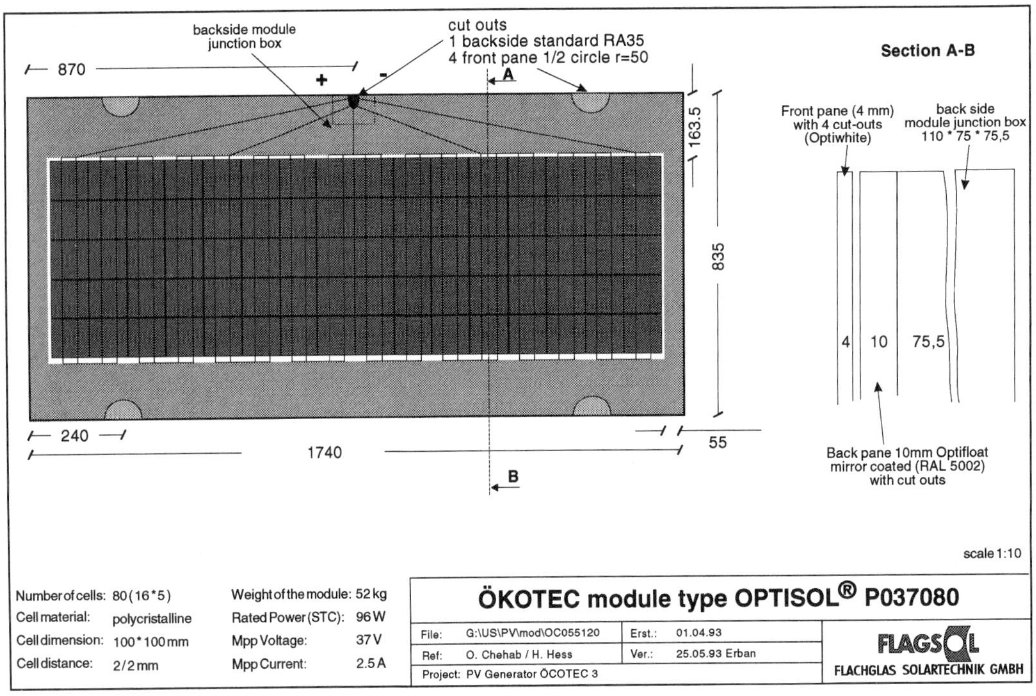


*Figure 14.4.3 Module design drawing.*

## 14.5 LRE building EPFL, Lausanne, Switzerland

*Figure 14.5.1 General view.*

**Key Data**

| | |
|---|---|
| Nominal power: | 3 kW |
| Gross area: | 28.5 m$^2$ |
| Module manufacturer: | Flagsol AG |
| Inverter: | Top Class 3000 |
| Start of operation: | 1993 |
| Owner: | EPFL Lausanne |

**Short description**

On the outer wall of the Electric Power Systems Laboratory the metallic thermo-lacquered cladding was replaced by photovoltaic panels in sheeted glass.

As existing elements were replaced, particular attention was given to the following details:

- identical dimensions through made-to-measure photovoltaic modules;
- uniform colour due to the use of polycrystalline cells aligned edge to edge without visible electrical connections; within the panel itself a frame painted the same blue as the metallic structure to hide the main electric connections;
- minimum specularity due to acid treatment of the glass (reduced reflection) to produce an effect as similar as possible to that of the former cladding and differentiate the facade elements from the windows.

The panels were mounted in a traditional way by VEC-glueing and recessed fitting of the metal sections.

**Mounting technique**

As can be seen in Figure 14.5.2 the mounting technique is quite traditional for glass panels:

- a metal frame is anchored into the wall, in this case using existing fixing points (Figure 14.5.3);
- two Z-shaped aluminium profiles are screwed onto the frame, over a protective sheet for the thermal insulation (Figure 14.5.4);
- the PV panel is slid into place onto the Z-shaped profiles.

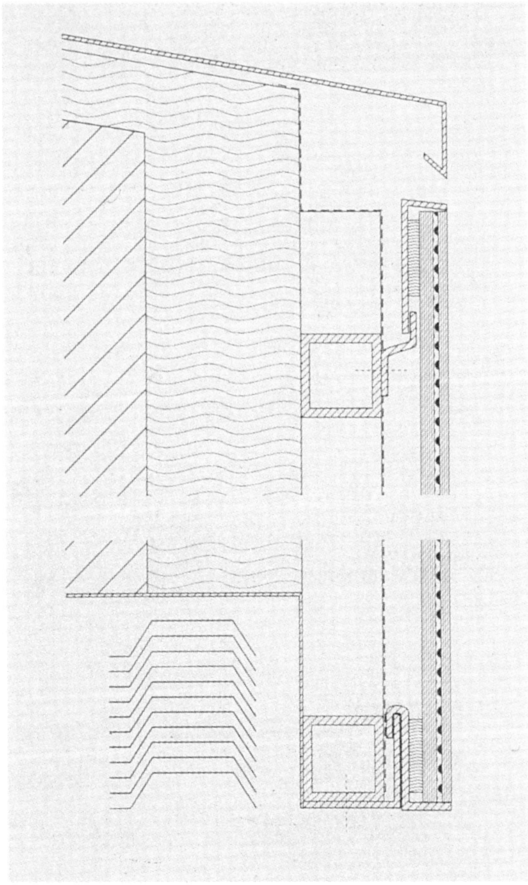

*Figure 14.5.2* *Cross-section of mounting design.*

*Figure 14.5.3* *Detail: Metal frame.*

*Figure 14.5.4* *Detail: Z-shaped profile.*

## 14.6 Scheidegger Metallbau, Kirchberg, Switzerland

*Figure 14.6.1 General view (see also colour plate 8).*

*Figure14.6.2 Detail of mounting structure.*

### Key data

| | |
|---|---|
| Nominal power: | 18 kW |
| Gross area: | 170 m² |
| Module manufacturer: | Solution |
| Inverter: | Solar Max |
| Start of operation: | 1992 |
| Owner: | Scheidegger Metallbau, AG |

### Short description

Just as important as the energy production here was the ability to gain knowledge about new architectural possibilities in facade design with PV modules. Besides the production of electricity and the aesthetic design, the weather protection of the building, the shading of workstations near windows in summer and the combination of the production of electricity and hot air are other goals of this installation.

To make a better use of the PV modules, a part of the facade is built with reflecting elements. The whole installation is working well, but there is heavy expenditure for attachment. Cheaper PV modules could be used effectively for the whole facade area.

The multifunctional design of the facade is shown in Figure 14.6.3.

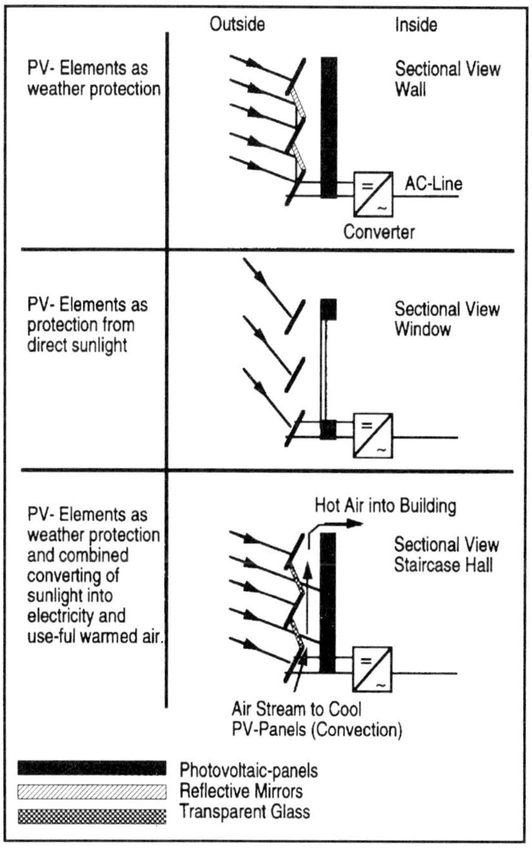


*Figure 14.6.3 Multifunctional design of the PV facade.*

## 14.7 House Weiss, Gleisdorf, Austria

*Figure 14.7.1 General view.*

*Figure 14.7.2 Detail of mounting structure.*

**Key Data**

| | |
|---|---|
| Nominal power: | 2 kW |
| Gross area: | 18 $m^2$ |
| Module manufacturer: | Siemens M50L |
| Inverter: | Siemens PV V 2500 |
| Start of operation: | 1992 |
| Owner: | private owner |

**Short description**

Within the framework of the Austrian '200 kW Rooftop Programme' Mr. Weiss installed a photovoltaic system on the roof of a building near his low energy house in 1992. The single family house has a living area of 150 $m^2$. It is located close to the small town of Gleisdorf in the Southern Austria. The heating load of the well insulated building was calculated to be 7 kW.

Mr. Weiss and his family already use very energy efficient household appliances. Data for recent years indicate, that only 2135 kWh of electric energy is annually consumed. This can only be achieved by using an 8 $m^2$ solar thermal collector system providing hot water for the dish washer and the washing machine. An auxiliary electric heating element is used to supply domestic hot water in wintertime (it consumes 335 kWh annually). The house is heated by a wood stove and some passive solar gains are utilized.

For better integration of the solar modules in the wooden roof structure the house owner decided to use Siemens M50L laminates. The solar generator consists of four strings with 10 modules each.

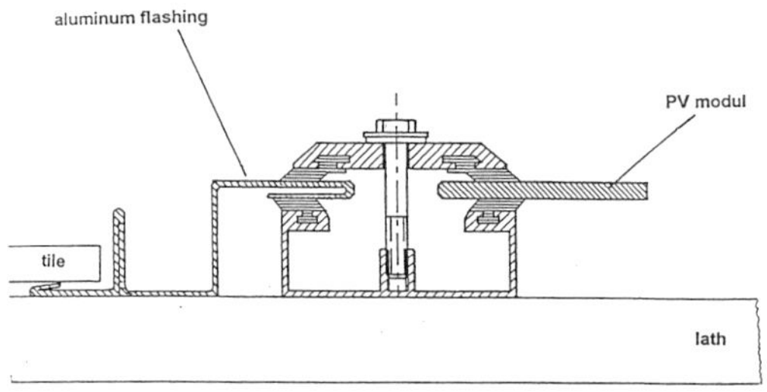


*Figure 14.7.3 Sectional view of the horizontal structure.*

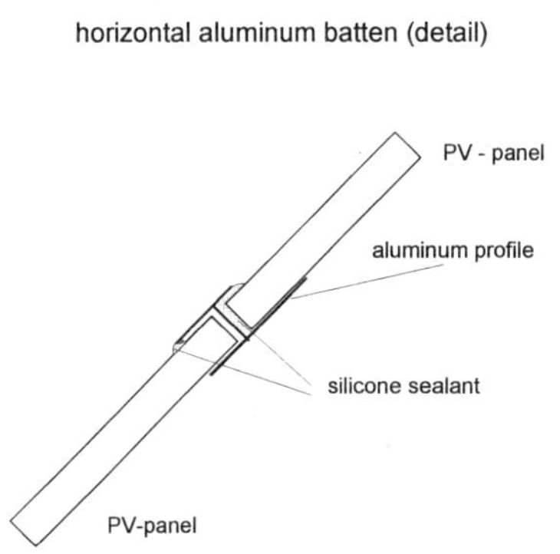


*Figure 14.7.4 Sectional view of the vertical structure.*

## 14.8 BOALsolar profile system for PV laminates

*Figure 14.8.1 Detail of mounting structure (R&S, The Netherlands).*

*Figure 14.8.2 Detail (R&S, The Netherlands).*

**Short description**

This profile has been developed by R&S, Ecofys and Boal (a large manufacturer of profiles for greenhouse constructions). The main objective of the system is to provide a standard PV mounting system for slanted roofs.

The profile has been applied successfully in a number of projects, including the 250 kWp roof integrated PV system in Amsterdam, The Netherlands, and a 100 kWp roof integrated system in Amersfoort (both in cooperation with the local utility).

The mounting system has been tested in accordance to general Dutch standards for roofing materials (including water tightness, robustness and fire protection).

The system consists of vertical and horizontal aluminium profiles. The vertical profiles are screwed to the battens (which are slightly higher than standard battens, in order to provide enough ventilation). Horizontal profiles are glued to the PV laminate (prefab) and fitted into notches in the verticals.

A plastic profile, placed over the vertical profiles, holds the laminates down and provides absolute water tightness.

Compared to mounting PV modules above the tiles, this system offers the following advantages:

- avoidance of tile and tiling costs;
- high aesthetic value;
- easy mounting technique;
- well-proven construction system;
- highly standardized, leading to lower costs;
- in-factory preparation of mounting (glueing of profiles).

*Figure 14.8.3 General view (Ecofys) (see also colour plate 9).*

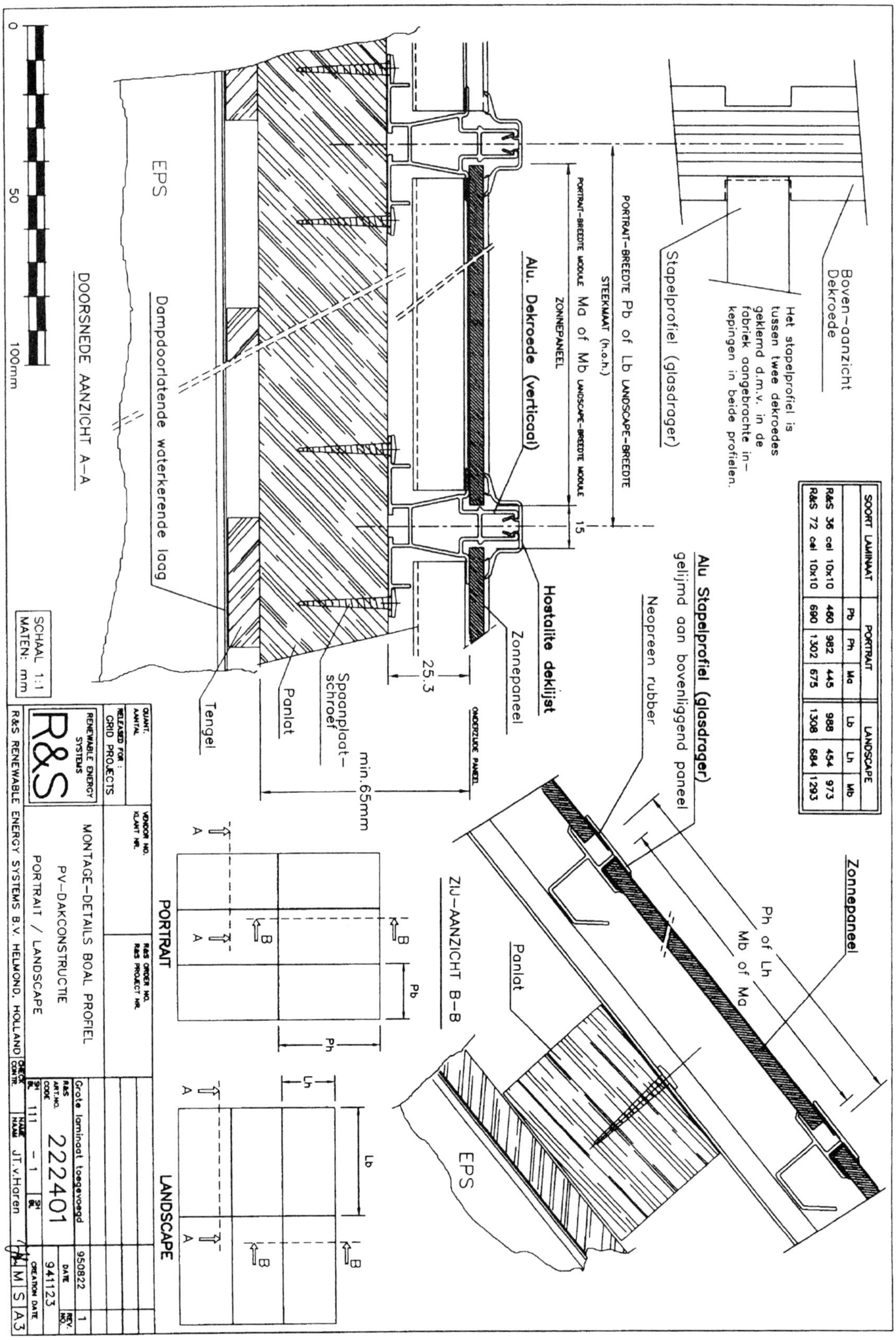


*Figure 14.8.4 View, front view, side view (© R&S Renewable Energy Systems, Holland).*

## 14.9 Solar tile, Mönchaltorf, Switzerland

*Figure 14.9.1 General view.*

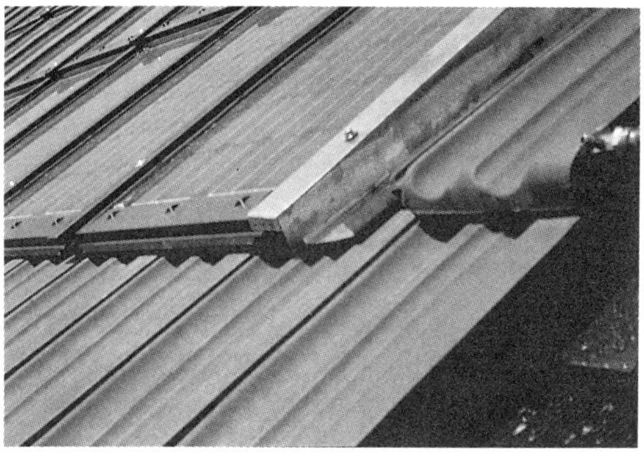

*Figure 14.9.2 Detail of mounting structure.*

### Key data

| | |
|---|---|
| Nominal power: | 3 kW |
| Gross area: | 36 $m^2$ |
| Module manufacturer: | Plaston-Newtec |
| Inverter: | Solcon 3300 |
| Start of operation: | 1992 |
| Owner: | PMS Energie AG |

### Short description

New solar roof tiles represent a real photovoltaic integration: the system is not installed on the roof; rather, the roof itself is the system. The roof elements installed on an old residential house in Mönchaltorf, made of glass, were developed and produced in Switzerland.

Each one of the 76.6 x 50.5 $cm^2$ large, rain proof resin boards with integrated photovoltaic cells replaces five conventional tiles.

Under the guidance of Alpha Real AG, a group of three companies has developed a solar tile to cover inclined roofs in one step. The tile is based on a standard laminating technology equipped with a special frame and newly designed module interconnections. Particularly, this new solar tile design had to employ the following functions:

- easy mounting like ordinary clay tiles;
- the essential function of the clay tile (i.e. weatherproof building material) has to be substituted by the solar tile;
- the sealing of the roof has to employ age old roof sealing techniques, as they are functioning on clay tiles without silicon or rubber sealant;
- the interconnection among solar tiles and between solar tiles and clay tiles has to be simple, easy to install, and absolutely reliable;

- to speed up installation time, the electrical connection from module to module has to be made with plug connectors;
- roofing professionals have to be able to do the installation and wiring work on the roof without extended training;
- every solar tile has to be easily exchangeable;
- the life assessment of the plastic frame has to be at least as long as the minimal expected life time of a solar module;
- it has to incorporate existing frameless solar modules of different thickness and to be able to adapt to future solar module development such as thin film technology or similar.

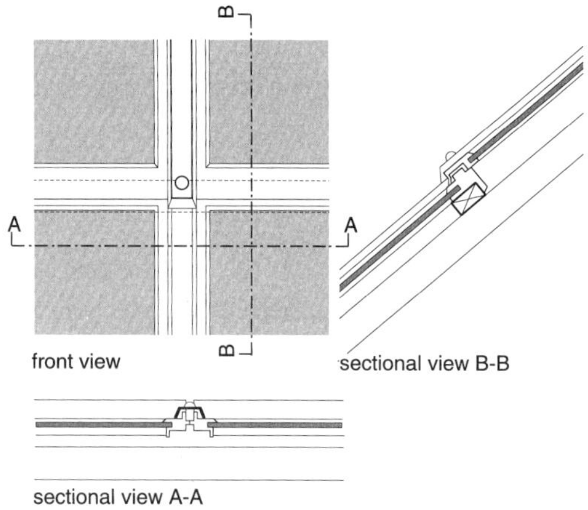


*Figure 14.9.3 Front view, sectional views.*

## 14.10 Pietarsaari Solar House, Finland

*Figure 14.10.1 General view (Neste Advanced Power Systems NAPS) (see also colour plate 10).*

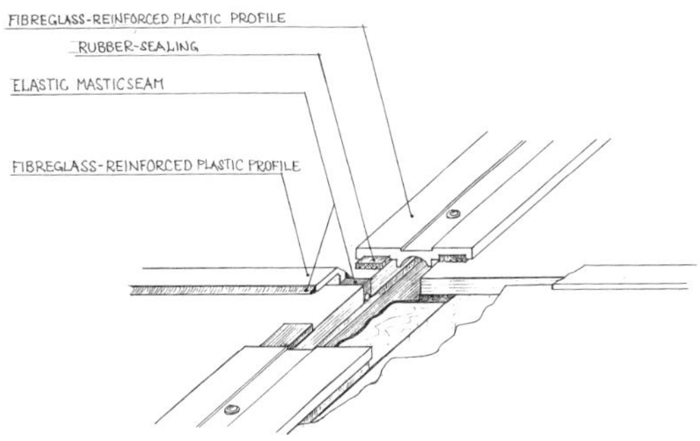


*Figure 14.10.2 Construction detail (Neste Advanced Power Systems NAPS)l.*

**Key data**

| | |
|---|---|
| Nominal power: | 2.2 kW |
| Gross area: | 55 m$^2$ |
| Module manufacturer: | APS |
| Inverter: | SMA 1800 |
| System design: | NAPS |
| Start of operation: | 1994 |
| Owner: | private owner |

The aim of the Pietarsaari Solar House, constructed in the Pietarsaari House Fair 94 area, is to demonstrate to a wide audience that with current commercial technology, energy consumption in houses can be dramatically reduced. During a House Fair exhibition more than 100,000 people visited the Solar House. In contrast to a conventional average Finnish house, which needs purchased energy of 160-250 kWh/m$^2$ for space heating, water heating and electricity, the corresponding energy demand has been reduced to roughly 20-30 kWh/m$^2$ in the Pietarsaari low energy house. The low energy consumption has been achieved by a combination of improved insulation, heat recovery from ventilation, heat pump with ground pipes, 'super-windows' as well as by utilizing solar energy. The house has 10 m$^2$ solar thermal collectors and a 2.2 kW grid-connected PV system.

The aim of PV integration into this house was to develop and test a simple and inexpensive watertight photovoltaic roof structure and to test it. The system has been constructed by using frameless large area amorphous photovoltaic modules and glass fibre profiles.

Amorphous modules were chosen for the ability to cover the whole roof with the chosen 2.2 kW array and thus to achieve a uniform appearance.

By using large area modules the number of connections and mounting profiles could be minimized. Also the short energy payback time of amorphous technology was considered to be significant. The benefits of composite profiles are:

- similar thermal expansion to that of glass;
- good electrical and heat insulation characteristics;
- good corrosion resistance and
- good mechanical strength.

The watertight PV roof was achieved by mounting the frameless modules over the existing vertical wooden roof substructure with rubber seals and long vertical glass fibre profiles going from the top of the roof down to the bottom of the roof. The horizontal seam between the modules has been sealed with a fork-shaped profile and elastic silicon mass. This type of simple integration method was planned in order to reduce the cost of material as well as the installation time.

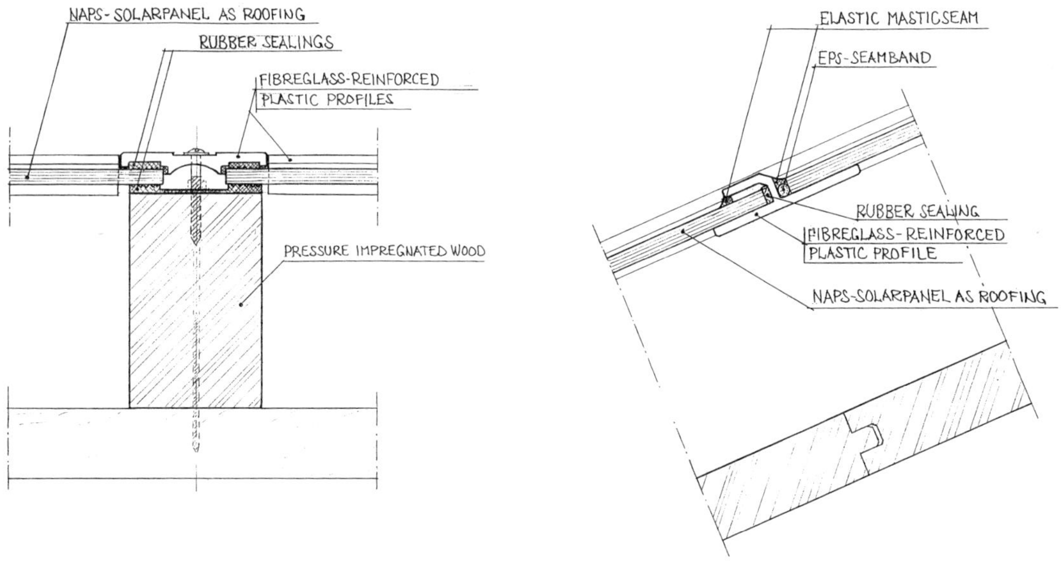


*Figure 14.10.3 Detail of mounting structure (Neste Advanced Power Systems NAPS).*

## 14.11 Office Building, Utility of Meilen, Switzerland

*Figure 14.11.1 General view.*

*Figure 14.11.2 Detail of mounting structure.*

**Key data**

| | |
|---|---|
| Nominal power: | 3 kW |
| Gross area: | 30 m$^2$ |
| Module manufacturer: | Helios |
| Inverter: | Solcon |
| Start of operation: | 1992 |
| Owner: | Utility Meilen |

**Short description**

Schweizer Metallbau AG has developed a roof-integrated system that is able to carry PV modules or thermal collectors. The installation area is completely covered by the construction, conventional tiles are only necessary to cover the remaining area of the roof. A combination of photovoltaic and thermal use of solar energy is possible.

A relatively large space on the back side of the module allows sufficient air circulation for module cooling and/or use of the low temperature heat. All current dimensions of PV modules or laminates can be applied, but there are some restrictions regarding the modules' edges. In order not to cover solar cells by rubber profiles, the distance between the edges and the first cells should be at least 15 mm. The thickness of the module in the edge zone must be within 4 to 5 mm.

The laminate is completely held by rubber profiles. There is no direct join to the aluminium profile. The top rubber profile is made of EPDM.

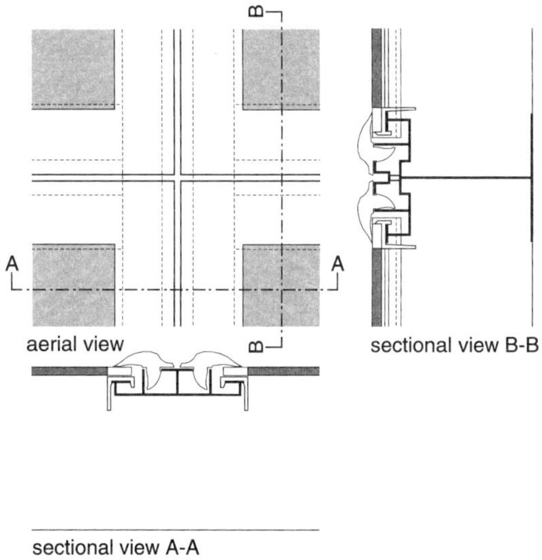


*Figure 14.11.3 Aerial view, sectional views.*

## 14.12 Integrated Solar Roof, Boston, USA

*Figure 14.12.1 General view (see also colour plate 11).*

*Figure 14.12.2 Detail of mounting.*

**Key data**

| | |
|---|---|
| Nominal power: | 4.3 kW |
| Gross area: | 40 m$^2$ |
| Module manufacturer: | Mobil Solar |
| Inverter: | American Power Conversion |
| Start of operation: | 1984 |
| Owner: | private owner |

**Short description**

In 1984, 10 million US viewers watched a TV programme introducing the latest technology in the area of photovoltaics at that time; it showed this one-family house in the north east of the USA. The roof facing south is tilted by 45° and divided into three sections. Twelve photovoltaic modules are located in each of the left and right sections, while the middle section contains roof-integrated solar collectors.

The well-insulated house was designed as a model for a country-wide energy-saving programme. The private residence meets high demands in terms of both aesthetics and comfort. The frameless PV modules are the finished weathering skin and are mounted directly over the wooden roof trusses without structural roof sheathing or additional weatherseal. Free air circulation behind the modules is encouraged with generous inlet and outlet vents designed to achieve maximum air flow. The PV laminates were designed specially for direct roof integration and the thermal collector uses the same glass cover. The result is a fully integrated appearance for both systems.

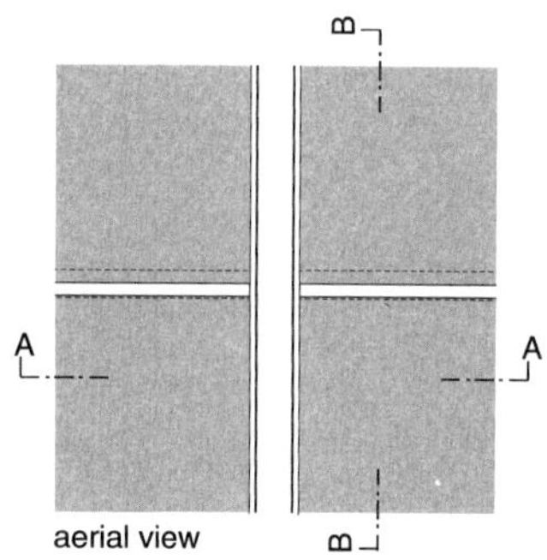


*Figure 14.12.3: Aerial view.*

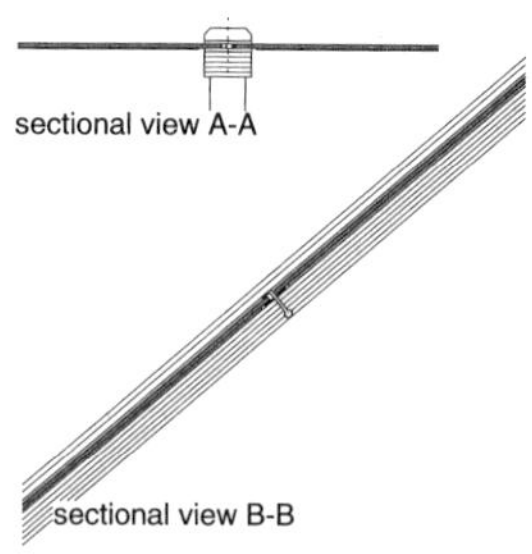


*Figure 14.12.4: Sectional views.*

*(All pictures: Courtesy of Solar Design Associates, Inc., Architects and Engineers, Harvard, MA USA.)*

## 14.13 Norwegian Solar Low Energy Dwelling

*Figure 14.13.1 General view (see also colour plate 12).*

*Figure 14.13.2 Detail of PV modules mounted on the roof.*

**Key data**

| | |
|---|---|
| Nominal power: | 2.52 kW |
| Gross area: | 22.6 $m^2$ |
| Module manufacture: | BP Solar |
| Inverter: | SMA PV-WR 1800 |
| Start of operation: | 1994 |
| Owner: | Hamar Building Society |

**Short description**

As this was the first grid-connected, roof integrated PV system in Norway; there were no ready-to-use mounting structures for PV modules on the market. Therefore, a safe and simple method of integration was adopted.

The PV panels were fitted on the south facing roof of the middle apartment of the three-unit row house. Prior to being mounted on the roof, the PV modules were fixed onto aluminium profiles in units of two, as shown in Figure 14.13.3. In this way, a quick and simple mounting procedure was obtained.

Aluminium profiles were placed in between the PV modules to prevent the accumulation of snow. This solution is not watertight, and therefore requires a complete, finished roof underneath. An alternative solution is to let the PV panels replace the conventional roof tiles. Thus, one would simply need a conventional sub-roofconstruction.

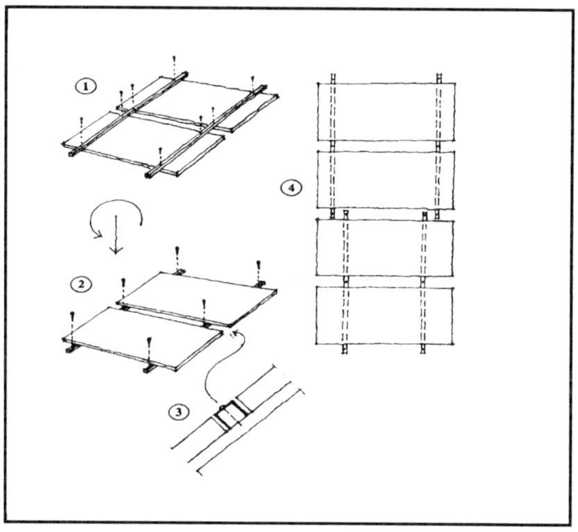

*Figure 14.13.3 Mounting procedure.*

## 14.14 Single Family House, Schauenbergstrasse, Zürich, Switzerland

*Figure 14.14.1 General view.*

*Figure 14.14.2 Detail of mounting structure.*

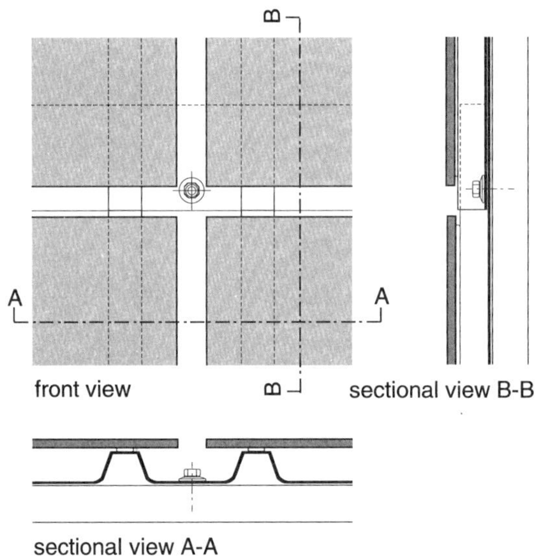


*Figure 14.14.3 Front view, sectional views.*

**Key data**

| | |
|---|---|
| Nominal power: | 3 kW |
| Gross area: | 25 $m^2$ |
| Module manufacturer: | Siemens |
| Inverter: | Solcon |
| Start of operation: | 1990 |
| Owner: | private owner |

**Short description**

The applied mounting technology can be adapted for facades. The use of trapezoidal sheet metal with modules glued on leads to a low cost, safe and flexible installation.

**Detail of mounting structure**

As glueing technique, a special rubber tape (acrylic foam) is applied. Glueing is a competitive mounting technique compared to mechanical attachment. Glueing does not penetrate the material nor causes it corrosion.

## 14.15 Shopping Centre, Stockholm

*Figure 14.15.1 General view.*

*Figure 14.15.2 View of the PV generator.*

**Key data**

| | |
|---|---|
| Nominal power: | 9.9 kW |
| Gross area: | 76 $m^2$ |
| Modulel manufacturer: | GPV Sweden |
| System design: | NAPS |
| Inverter: | 2 x Solcon 3400 HE |
| | 2 x PV-WR 1800 |
| Start of operation: | 1993 |
| Owner: | Stockholm Energi AB |

**Short description**

This photovoltaic project was the contribution from Stockholm Energi AB to the Swedish participation in Task 16. The system is also the Swedish 'Demobuilding' and the largest PV installation in Sweden. The 90 modules are divided into three separate systems, each equipped with its own inverter. Two of the systems are mounted on the 45° tilted roof of a shopping centre in Stockholm City. The third part is installed on the vertical wall of a roof space on top of the building. The modules are standard GPV110, framed modules, with black Tedlar™ on the back side to make the array visually well integrated into the traditional black roofing sheet.

One of the tilted arrays and the vertical one are connected to Solcon inverters. The other tilted system is connected to two PV-WR1800 inverters operating in a master-slave configuration.

**Detail of mounting structure**

The modules are mounted in groups of three into a framework of black anodised aluminium U-profiles that are fixed to the roof or the wall in the case of the vertical array. A cable duct is attached to one of the horizontal profiles. Each module is lifted up in the upper U-profile and then slid into the lower U-profile. By tightening screws on both the upper and the lower profile the modules are fixed into the structure. This method makes it very easy to remove any module for service. The cabling was laid out in the cable duct before installing the modules. The modules were delivered with cables and quick-coupling connectors which made the electrical installation work safe and relatively fast. The connectors were then sealed by shrinking tubing.

*Figure 14.15.3 The upper part of the mounting frame.*

*Figure 14.15.4 The lower part of the mounting frame with the cable duct attached below.*

## 14.17 Solarzentrum Freiburg, Germany

*Figure 14.17.1 General view (see also colour plate 13).*

*Figure 14.17.2 Facade detail.*

**Key data**

| | |
|---|---|
| Nominal power: | 18.5 kW |
| Gross area: | 153 m$^2$ |
| Module manufacturer: | DASA |
| Inverter: | SKN402 /SKN 301 |
| Start of operation: | 1993 |
| Owner: | SST GmbH |

**Short description**

Four PV generators are integrated into the facade and roof of this office building. The facade is of special interest, since it is a structural glazing (SG) facade.

SG is a construction method using a special type of silicone adhesive to glue window and parapet elements, also made from glass, directly onto the statically bearing support structure. Additional fixtures, which are possibly visible from the outside are in many cases not necessary. The silicone adhesive performs sealing function and it takes all mounting forces as well as wind loads, transferring these to the support structure. The support structure consists of thermally decoupled aluminium profiles, which are anchored to the building at the various storey ceilings.

Figures 14.17.3 and 14.17.4 show details of the sophisticated support structure, which have to fulfil very different functions: static support, thermal decoupling, accommodation of window elements, accomodation of parapet elements and accommodation of regular wall insulation.

An SG facade is turned into a PV facade by replacing the glazing material by customized frameless PV modules. The whole facade is divided into segments. Each segment, which may reach up to 8m height, is completely factory-mounted including thermal insulation and a lining which eventually makes the inner wall cover. PV modules, regular wall elements as well as windows, can all be mounted in one segment.

Factory preconstruction of segments including PV modules offers a high production quality at very low tolerances. Thus cable glands etc. can be placed exactly in the area of installation channels or hollow ceilings. Preconstruction minimizes the risk of damaging the valuable PV modules in a harsh construction site environment.

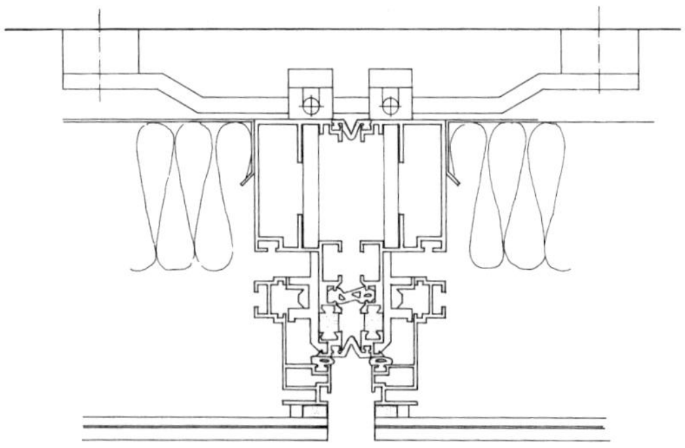

*Figure 14.17.3 Horizontal cut through facade construction.*

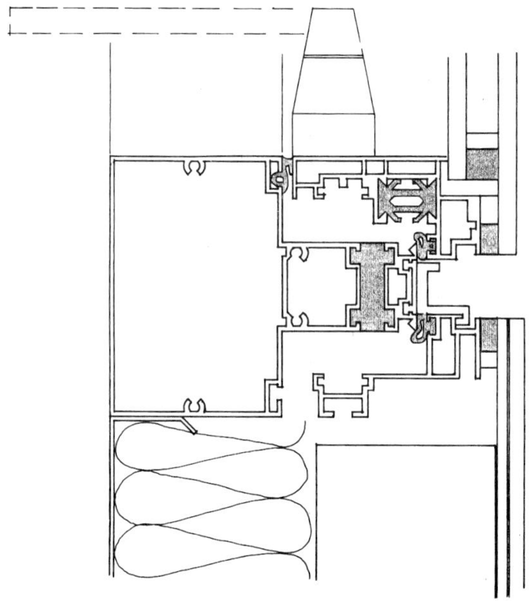

*Figure 14.17.4 Vertical cut through facade construction.*

## 14.18 SOFREL, Solar Flat Roof Element

*Figure 14.18.1 General view (see also colour plate 16).*

*Figure 14.18.2 Detail of mounting structure.*

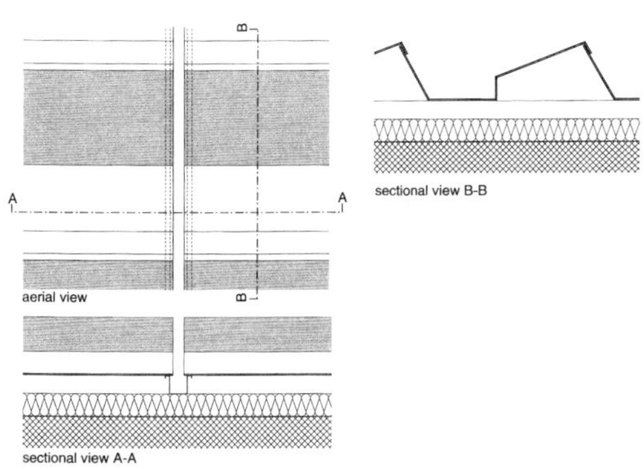


*Figure 14.18.3 Aerial view, sectional views of a SOFREL® with sheet metal or a composite material such as Alucobond®.*

**Short description**

SOFREL® (Solar Flat Roof Elements) is a flat roof integrated PV system.

This new project has recently started in Switzerland which aims to develop a Solar Flat Roof Element (SOFREL®). PMS Energy Ltd., Alpha Real Ltd., Swiss Institute of Technology in Lausanne (EPFL) and the Union Bank of Switzerland (SBG/UBS) are developing a flat roof integrated PV system. The main goal is to combine the function as building skin and PV element in one unit for flat or nearly flat roofs.

As shown in Figure 14.18.3 the solar modules are integrated in composite materials such as Alucobond® which results in a watertight roof together with the drainage system. The difference between watertight and non-watertight SOFREL® is explained Figures 14.18.5 and 14.18.6.

Non-watertight versions are made of steel reinforced concrete. In such a case, the flat roof is sealed with a conventional method.

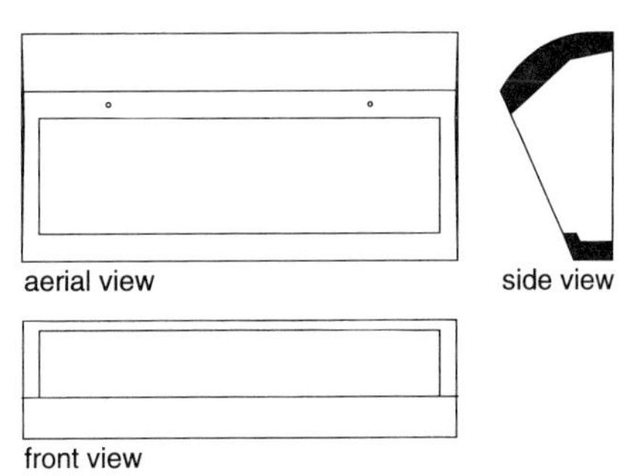


*Figure 14.18.4 Aerial view, front view, side view of a SOFREL® element based on concrete.*

The SOFREL® system began to enter the market during 1995. A patent is pending; produc-

## Chapter 15

# Design Considerations

### 15.1 Overview of the design process

This section gives simple methods and tools of PV system design for a non-PV specialist. It is recommended, however, that the detailed final system design should be done by a specialist because of liability and responsibility matters, as well as technical familiarity.

The design process of a PV system is relatively straightforward. The design starts by screening the site and the building in question to evaluate the applicability of PV for electricity production in the specific case. This means that the solar availability must be evaluated for the region and building in question. Also, the energy efficiency of the loads should be considered. The trade-off questions between solar thermal and photovoltaics must be considered especially if there is lack of suitable building surface area. It should be noted that although solar thermal energy is cheaper, it is of little value, if the user needs electricity. In Appendix VI this topic is discussed in more detail.

After the decision on the suitability of PV has been made, the design starts with a detailed load analysis (Chapter 16) in order to improve the electricity utilization efficiency of the building and to get the basis for the PV system sizing. Following this, a rough system sizing can be performed (Chapter 17) giving the boundaries for the PV system component selection (Chapter 18). Then the house owner can turn to the PV system suppliers for a detailed offer for the system he needs. The design process described above is illustrated in Figure 15.1.

There are also several general institutional matters affecting the PV system, which should be considered at certain stages of the system design. These matters include: the utility interface and interconnection issues, land use and construction regulations, safety, financing, liability and insurance issues.

### 15.2 Screening the application site and the building

The general site aspects affecting the system design are: the actual location of the system, available array area, solar access to the considered PV array surface and the energy efficiency of the building.

The **location** of the system mainly affects the **amount and the value of the PV energy**. The value of the PV system is determined by the amount of energy produced annually and the value of that energy. The annual energy production depends proportionally on the incident radiation. The value of PV energy depends on the local utility's avoided costs and the utility's pricing policy concerning electricity buyback in grid-connected systems. In stand-alone systems this depends on the price and reliability of the alternative electricity production possibilities. Both insolation and PV energy value will vary from region to region. Also architectural, aesthetic and environmental matters can increase the value of a PV system.

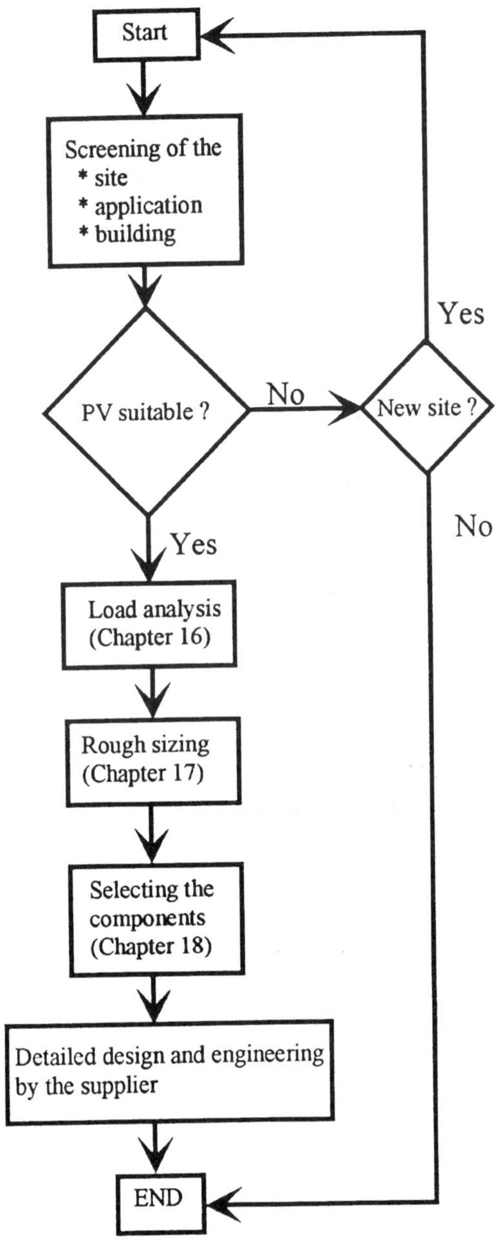


*Figure 15.1 Design process.*

The **available roof or facade area** facing approximately south (in the northern hemisphere), specifies the **maximum PV array size**. Today's commercial thin-film modules need 20-30 $m^2$ for a kW and crystalline modules need 8-15 $m^2$. In grid-connected systems the **minimum system size** is about 1 kW, because the price per kW increases rapidly below that range. This is due to a significant cost for size independent wiring, power conditioning and engineering. However, with the development of module-integrated inverters this minimum practically vanishes. In stand-alone systems there is no minimum size, because there is no grid electricity available and the minimum size is given by the load size.

**System performance** is mainly affected by solar access, which has to be guaranteed both at the time of system construction and also in the future. In general, no roof structure (chimneys, offsets, projections, antennae) or surrounding objects like trees and other buildings should shade the PV array at any time of the year. Shadowing - even on small parts of the array - cuts down the PV electricity produced and may also lead to PV array matching problems. Small soft shadows of distant objects are not a concern, but dark large shadows should be avoided. If solar access cannot be guaranteed now and in the future, the application site is not suitable for PV systems. Figure 15.2 illustrates solar access by two examples.

The energy generated by a PV system can have a high value in favourable applications. However, if the building has not been designed to be **energy-efficient,** it is a waste of valuable photovoltaic energy (and building owner's money) to power inefficient appliances. Therefore, before designing a PV system, the house loads should be decreased by using the best commercially available energy efficient appliances (Chapter 16).

### 15.3 Utility interconnection issues

Conventional energy production in industrialized countries is based on centralized power stations. Photovoltaic systems, and especially PV systems integrated into buildings, are very small compared to these. Utility companies in many countries are often not familiar with this kind of decentralized energy production form. Thus the utility interconnection, which is necessary for grid-connected systems, has been a barrier for the PV system implementation and

*Plate 1* *Handelshaus A. Wild, Innsbruck, Austria*
*13 $kW_p$ photovoltaic facade (IEA Task 16 Demonstration Building).*

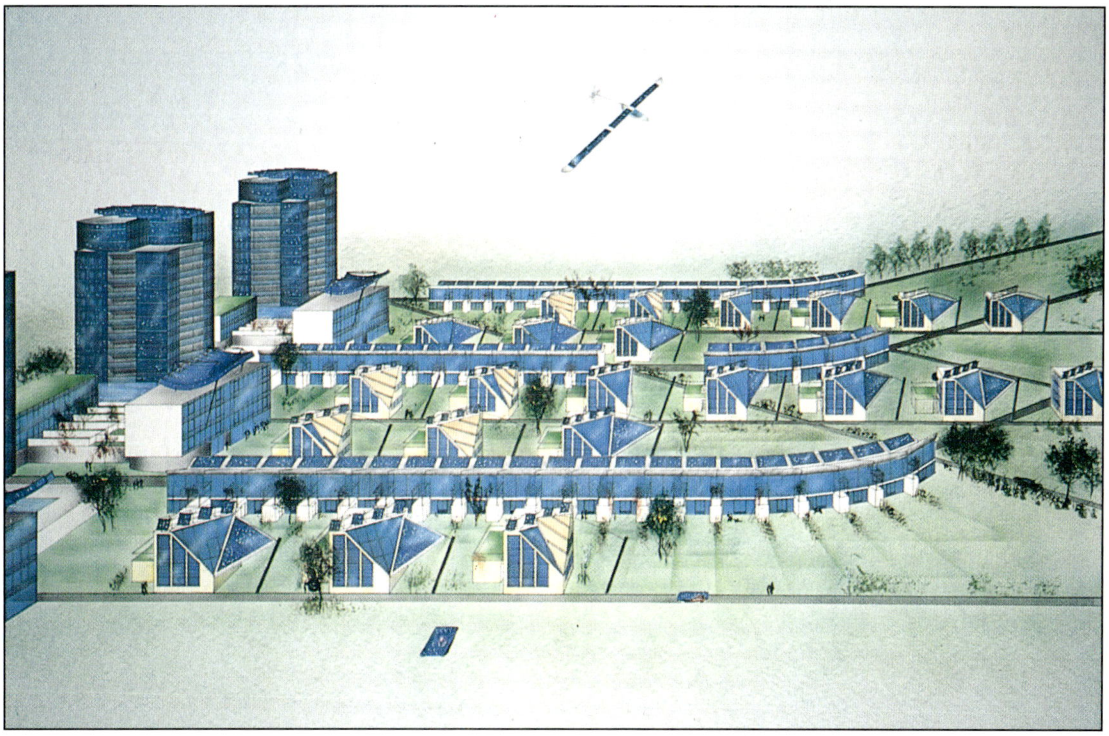

*Plate 2* *IEA Task 16 Architectural Ideas Competition*
*Entry by B.J. van den Brink, The Netherlands*
*(see also Figure 13.16, page 106).*

*Plate 3* *Church, Steckborn, Switzerland, 18.6 $kW_p$ facade-integrated PV system (see also Figure 11.2, page 78).*

*Plate 4* *Stadtwerke Aachen, Aachen, Germany, General view (see also Figure 14.2.1, page 118).*

*Plate 5* *German Architectural Ideas Competition 1994, "Zentrum für Kunst und Medientechnologie", Karlsruhe, Germany*
*Visible photovoltaic panels above the entry hall (see Fig. 13.23, page 110).*

*Plate 6* *Ökotec, Office Building, Berlin, Germany*
*General view (see also Figure 14.4.1, page 121).*

*Plate 7* *Ökotec 3, Office Building, Berlin, Germany*
*Detail of PV element (see also Figure 14.4.2, page 121).*

*Plate 8 Scheidegger Metallbau, Kirchberg, Switzerland General view (see also Figure 14.6.1, page 125).*

*Plate 9 Energy-autonomous house at Woubrugge, The Netherlands (Ecofys) BOALsolar profile system (see also Figure 14.8.3, page 127).*

*Plate 10* *Pietarsaari Solar House, Finland*
*General view (NAPS) (see also Figure 14.10.1, page 131).*

*Plate 11* *Integrated Solar Roof, Boston, USA*
*General view (see also Figure 14.12.1, page 134).*

*Plate 12* *Solar Low Energy Dwelling, Hamar, Norway*
*General view (see also Figure 14.13.1, page 135).*

*Plate 13* *Solarzentrum Freiburg, Germany*
*General view (SST GmbH) (see also Figure 14.17.1, page 141).*

*Plate 14* *Demosite, Lausanne, Switzerland*
*General view (see also Figure 14.19.1, page 145).*

*Plate 15* *Demosite, Lausanne, Switzerland*
*"Pyramides" combining sun protection, natural lighting and electricity production (see also Figure 14.19.4, page 146).*

*Plate 16* *SOFREL, Solar Flat Roof Element*
*General view (see also Figure 14.18.1, page 143).*

*Plate 17* *A photovoltaic and thermal co-generation facade system, Fondi (Latina), Italy*
*General view (see also Figure 14.21.1, page 148).*

still is in some countries. The problem has been addressed by many countries where the possibility of decentralized energy production is now guaranteed. There are, nevertheless, major differences between countries and also utilities based on the judgement of the utility management regarding the avoided costs, acceptable risk, possible equipment failure and the limited experience of the utility with PV system interconnection. Thus in case of grid-connected PV systems, the system designer should contact the local utility at the very beginning of the system design to establish the interconnection framework and the possibility of a contract between the utility and the PV system owner.

*Figure 15.2a* *Solar access to the PV array hindered by different obstructions.*

*Figure 15.2b* *Solar access to the PV array hindered by surrounding houses.*

## 15.4 Land use, design and construction regulations

Land use requirements exist to define acceptable standards of community development in the interest of public health, safety and welfare. Several requirements exist, such as zoning ordinances, restrictive covenants and subdivision statutes.

Several key categories of land use issues that can affect the implementation of residential PV systems include:

- the guarantee of solar access for the PV system;
- compliance with building and site location regulations;
- compliance with conforming use regulations;
- compliance with design, style, materials or colour regulations.

Of these issues, the guarantee of solar access for the PV system is the most critical, since legal precedent does not give the PV system owner any rights to the sunshine crossing adjacent property without a contract or ordinance.

Requirements such as standards of recommended design practice, testing standards and manufacturing standards establish a minimum level of quality and reliability in equipment, construction methods and materials. Because these standards are frequently referenced by construction codes, utility interconnection agreements, financing agreements or insurance agreements, they play a crucial role in PV system implementation.

## 15.5 Safety issues

Most safety concerns associated with residential PV systems can be addressed by careful attention to system design, i.e. good engineering practice. Major safety concerns arise from the fact that the sun cannot be turned off. Several less serious safety concerns are also associated with residential PV systems.

The following **checklist** serves as a reminder:

- PV-unique hazard potential understood by design team, installation crew and homeowner;
- safety procedures, based on good engineering practice, followed;
- safety devices employed;
- risks to unauthorized persons minimized;
- equipment stored in secure area;
- emergency operating conditions addressed in system design;
- service access that minimizes hazards provided;
- rapid, efficient snow shedding addressed.

## 15.6 Financing issues

Financing issues for residential PV systems are not very different from those for conventional systems in residences. However, incentives are created from time to time by local and governmental offices to encourage the utilization of renewable energies like photovoltaics. The aim of these incentives is to reduce the solar energy cost for the consumer, either by decreasing the investment cost or by decreasing the system operation cost. These kind of incentives include tax credits, exemptions from sales/use tax, grants and loans.

## 15.7 Liability and insurance issues

Reasonable care is required in the design, manufacture, assembly and operation of the PV system in order to protect it, the residence and the utility system from damage as well as any operating personnel from injury. Product liability and owner liability must be considered. Product liability concerns the strict liability, warranties, negligence and misrepresentation. Owner liability concerns the degree of protection afforded to persons and property.

The types of insurance to be considered are product liability for the manufacturer, professional liability for the designer and casualty and liability insurance for the homeowner. The potential hazards in dealing with PV systems are:

- PV system damage from environmental degradation;
- vandalism;
- personal injury to installer, occupants and others;
- property damage;
- fire damage;
- chemical damage (e.g. batteries);
- communications interference (e.g. inverter).

## Chapter 16

# Load Analysis

### 16.1 Identifying loads

The very first step in designing a PV system must be a careful examination of the electrical loads. The reasons are twofold:

- Obviously, the sizing of the system components is dependent on the electricity and power demand. For stand-alone systems, this is crucial.
- Oversized systems resulting from a poor load analysis and the idea of staying on the 'safe side' increase the system costs. This is particularly damaging in a field where poor economics are a major drawback, which still is the case for PV.

The second reason also leads to the important issue of minimizing loads without decreasing the user's comfort. This issue will be addressed in section 16.2.

The emphasis of this section is to give advice on how to estimate the actual loads. Past experience shows that small power consumers are often neglected or simply forgotten and add up to considerable loads in the end, e.g. the stand-by consumption of consumer electronics. A complete list of all possible electrical consumers would not fit into this design book. Table 16.1 should serve as a guide and reminder.

Both the rated power of the load and the energy demand are important for correct sizing. The energy is obtained by multiplying the power with the time of operation, called the duty cycle in Worksheet #2 (Appendix II).

- Lighting
- Clocks
- Pumps
- El. heaters
- Fans
- Air conditioners
- El. driven blinds
- El. driven doors (e.g. garage)
- Elevators
- Security systems
- Coffee machines
- Refrigerators
- Freezers
- Dishwashers
- Ovens
- Microwave ovens
- Toasters
- Grills
- Mixers, Blenders
- TV, VCR, Radio, Stereo
- Projectors
- Washing machines
- Irons
- Dryers
- Vacuum cleaners
- Hair dryers
- Shavers
- Sauna heaters
- PCs with Monitors
- Printers
- Copy machines
- Communication equipment (fax, phone)
- Other office equipment
- Battery chargers
- Monitoring equipment
- Tools
- Toys

*Table 16.1 Examples for electricity consumers.*

## 16.2 Improving energy efficiency

One of the most cost-effective sources of energy is 'saved energy'. In PV systems the replacement of loads or appliances by more efficient ones is often economic because the investment costs for the new items are lower than those for the PV system components (modules, batteries, larger-rated power conditioning equipment) required to provide the difference in energy consumption of the old and the new loads or appliances. If, for instance, the installed nominal power of a residential system could be reduced by 1 kW (worth about 12,000 US$ in grid-connected and 25,000 US$ in stand-alone systems!) due to a number of more efficient household appliances, it would easily pay for them.

It is therefore not only cost-effective but crucial to perform a careful examination of the electrical consumers. There are two levels of reducing the energy demand: first, one should eliminate all electrical consumers that are not necessary at all or ecologically disadvantageous. Second, for the remaining consumers the most efficient ones should be selected.

The first category usually comprises thermal equipment. As long as electricity is produced centrally with an efficiency of 30 to 33% while generating mostly waste heat, it is not reasonable to use this valuable electricity for heating purposes. Electric space heating should be avoided in most cases. Exceptions, for practical reasons, might be radiators in the bathroom operated only for a very limited time per day or similar applications, where the total energy demand is very low.

Water should be heated using conventional burners, possibly in addition to solar thermal collectors. Washing machines and dish washers can be equipped with hot water inlets. The electric heating coil then only compensates the temperature difference (if any) of the hot water and the set point.

Air-conditioning equipment is not always necessary. Especially in new buildings, with careful planning using expert knowledge and sophisticated computer simulation tools, it is often possible to avoid air-conditioning equipment. If there is still the need for air conditioning, intelligent controls and reasonable temperature set points can significantly reduce the loads. A fixed set point of 18°C to 20°C throughout the summer is a waste of energy. At outside temperatures of 35°C one feels comfortable at, say, 28°C.

The cooling load could also be minimized through solar control, which is obtained by careful design of the building and different kinds of shading devices. Utilization of natural daylight through building design could both reduce the electricity demand for lighting and thereby also the cooling load.

The heating and ventilation loads of the building should also be minimized. This could be achieved through different means, e.g. by using a high degree of thermal insulation of the envelope, good air tightness and efficient heating and ventilation systems.

After eliminating electrical consumers that should not be there or not be electrically driven, the second stage is to improve the energy-efficiency and the right choice of electrical consumers has to be considered. For a selection of household appliances, Table 16.2 gives the range of annual energy consumption from very efficient to standard to poor.

## 16.3 Load management

The sizing of an energy supply system depends not only on the expected energy consumption over a given period but also on the peak power demand. This is true for the big utility grids as well as for the owner of a small stand-alone PV system. To ensure supply reliability, the system must be able to meet the maximum desired peak loads. If this happens at night-time in a PV system, the battery and/or the back-up generator must do the entire job. It is obvious that a load profile following the supply curve of the incident radiation would be advantageous. This is where load management comes in. Along with the sensible selection and use of efficient home appliances, the practice of load management is a sound method of assuring that each kWh used is worth the money spent on it. At its simplest, load management consists of a series of conscious decisions concerning when to add certain appliances or power-consuming functions to the load. For instance, you may decide not to run the washing machine, vacuum cleaner and oven simultaneously. Equipment that does not have to be in operation at any given moment, as long as it runs for a certain amount of time over a given period, is well-suited for load management controllers. Air conditioning, refrigeration or well-water pumping equipment are examples. Load management microprocessors can be programmed to control loads according to user-specified strategies. These can be fixed-priority strategies, rotational strategies, where a number of loads are turned on and off in a sequential cycle, or combinations of both.

It will always be helpful to have the expected typical daily load curve at hand (see e.g. Figure 16.1). In existing households, it can be monitored. Otherwise it must be estimated. The next step is the analysis of sharp peaks. If they can be reduced or if parts of the load can easily be moved to periods with greater electricity production, a load management strategy should be applied.

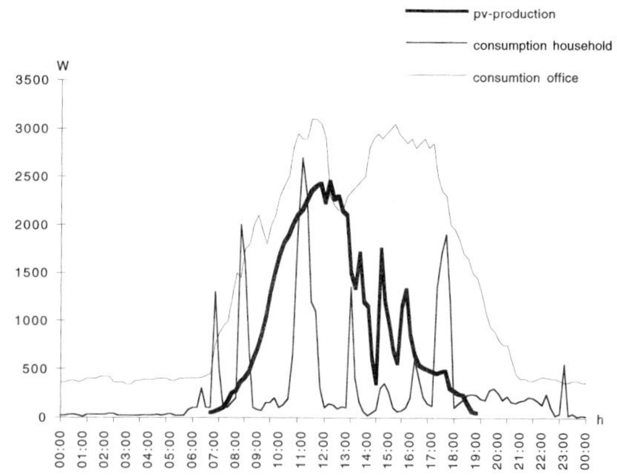


*Figure 16.1 Daily load profile of a 5-person household and an office with 10 work places compared to the energy production of a 3 kW system in partly overcast conditions.*

| | low | std. | high |
|---|---|---|---|
| Refrigerator[1] | 87 | 230 | 270 |
| Freezer[2] | 168 | 426 | 800 |
| Refrigerator/ Freezer[3] | 267 | 343 | 625 |
| Washing Machine[4] | 280 | 366 | 522 |
| Washing Machine[5] | 202 | -- | -- |
| Dish Washer[6] | 296 | 481 | 614 |
| Lighting[7] | 87 | -- | 438 |

1 200 l, no freezing compartment
2 200 l
3 200 l
4 without hot water inlet, 3 cycles per week
5 with 60 °C inlet, 3 cycles per week
6 without hot water inlet, 5 cycles per week
7 4500 lm, 4 hours per day

*Table 16.2 Electricity consumption ranges of typical electrical consumers [kWh/a].*

## Chapter 17

# System Sizing

### 17.1 Introduction

Before choosing the final components, the system should be roughly sized to allow viewing of approximate component sizes. Later, the components must be sized again by a detailed electrical and mechanical design. The purpose of this chapter is to provide **simple tools** to roughly estimate the needed system size before contacting a PV specialist.

### 17.2 Sizing procedure

In general PV systems in buildings are sized in such a way that the PV system can meet the building loads either fully or partially and still function reliably. In stand-alone and hybrid systems, the batteries and/or backup system (i.e.: diesel generator) must deliver the electricity even during long overcast periods. In grid-connected systems, there is no storage component because the grid acts as an infinite buffer.

The key factors affecting system sizing are the load size, the operation time (all year, summer only etc.), the location of the system (solar radiation) and a possible sizing safety margin. Besides that, the **available roof or facade area** can restrict the PV array size. Finally, the most important restriction for PV system sizing is the available budget. Roof/facade area and budget are typically the key restrictions for the design of a grid-connected PV house.

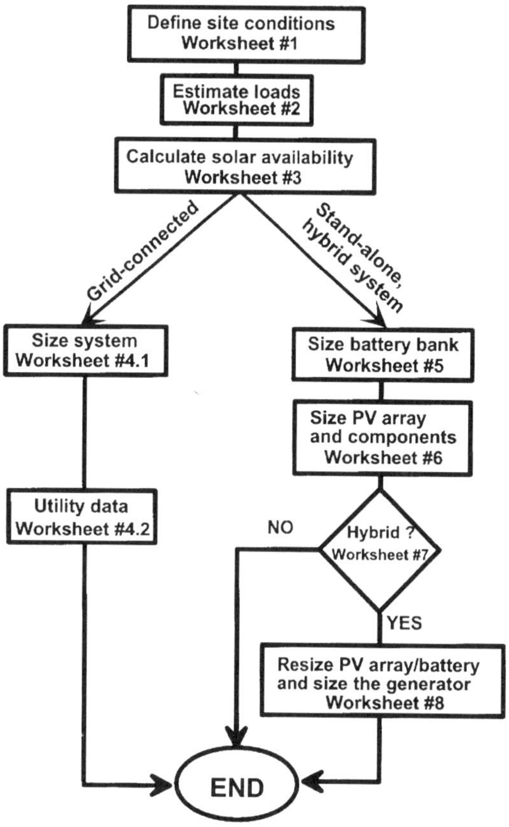


*Figure 17.1 Steps of the rough sizing procedure.*

Steps involved in the **rough sizing procedures** for different types of PV building systems are presented in Figure 17.1. The approach is to estimate the required component sizes by making assumptions about the efficiency of all key components and by using monthly average weather data. To make the procedure easier, a set of Worksheets (#1-#7) has been prepared for the different steps (see Appendix II).

**Specification of site conditions (Worksheet #1)**

Define site and weather station location (latitude, longitude) and monthly average values of the global irradiance on the horizontal surface (kWh/m$^2$) and the annual average as well as the minimum and maximum monthly average ambient temperatures. Weather data for several locations are available from Appendix I. The weather station chosen should belong to the same climate zone and it should be as close as possible to the site in question. This is especially important if the PV system site is close to a high mountain or a coast.

**Estimation of solar availability (Worksheet #1)**

Main factors affecting the solar availability are the orientation (tilt and azimuth angles) and the possible shading caused by the surroundings. By multiplying the horizontal insolation values with a monthly tilt and azimuth angle factor, the monthly radiation values on the module surface can be estimated. In Appendix I, this monthly factor is presented for different locations for horizon shadowing levels of 0, 20 and 40 degrees.

Ground reflection, shadowing from the neighbouring buildings and from the PV building itself can affect the solar availability as well. However, these effects are too complex to consider in this rough sizing procedure. The surface with the highest available radiation (kWh/m$^2$) during the system operation time should be chosen. If that surface is not large enough, other surfaces should be considered.

In Figure 2.5 the dependency of available solar radiation on tilt and orientation angles is shown for Central Europe. If these angles can be chosen freely (e.g. in a new building), the choice should be made on the basis of maximizing the PV production. The annual PV output is maximum when the azimuth angle is within $\pm 45°$ from the south orientation (in the northern hemisphere) and the tilt angle is $\pm 15°$ of the latitude angle value. Larger variations are not recommended. In practice, architectural and technical reasons usually limit the possible orientation.

**Estimation of the electricity demand (Worksheet #2)**

For an existing building, past electricity bills will help to perform this task. For a building not yet constructed, guidelines are given in Chapter 16 on how to estimate energy demand.

**Sizing of a grid-connected system (Worksheet #3 I&II)**

The optimum size of a grid-connected system also depends on a number of external factors such as: the investment cost of the system, the available budget, governmental subsidies, the energy payback policy of the local utility, and the amount of PV energy directly used by the building. It must be remembered that because of the variable nature of PV power it is seldom used to decrease the peak load demand of the building.

The buyback ratio is the major utility factor affecting the sizing of the PV system. This is the ratio between the price the utility pays for the PV electricity and the price of the electricity bought from the grid. Typically the buyback ratio is less than one (0.5-1.0), which means that the utility pays less for PV electricity than the building owner pays for the grid electricity. Therefore, most of the PV system production should be used directly in the building. This use depends on the matching of the PV production with the house load profile.

The shaded area in Figure 17.2 represents an average result obtained using a typical European and North American climate and load profile.

Typical sizes of a grid-connected system for a single family house range from 2 to 5 kW (with an annual electricity consumption of 4-5 MWh). This equates an annual electricity production of 1.5-3.8 MWh in Northern Europe, 1.6-4.0 MWh in Central Europe and 2-5 MWh in Japan, assuming optimum orientation and design.

In practice, the nominal size of the PV array should be chosen based on the load size and the budget. The rule of thumb is that an installed grid-connected PV system will cost 10 US$/$W_p$ (1994 price). The required PV module area $A_{PV}$ ($m^2$) can be calculated from the chosen nominal PV power using the formula

$$A_{PV} = \frac{P_{PV}}{\eta_{PV}}$$

where $P_{PV}$ (kW) is the nominal power of the PV array under standard test conditions (STC) and $\eta_{PV}$ (fraction) is the efficiency of the modules at STC (see Table 17.2).

The annual energy production of the system can be calculated using the formula below:

$$E_{PV} = \eta_{BOS} K_{PV} P_{PV} S$$

where S ($kWh/m^2$) is the annual solar radiation on the PV array, $K_{PV}$ is a decreasing factor (~0.9), which takes into account phenomena such as module temperature, dust, array imbalance, circuit losses etc. and $\eta_{BOS}$ is the balance of system efficiency which is the system efficiency without the PV module efficiency.[1]

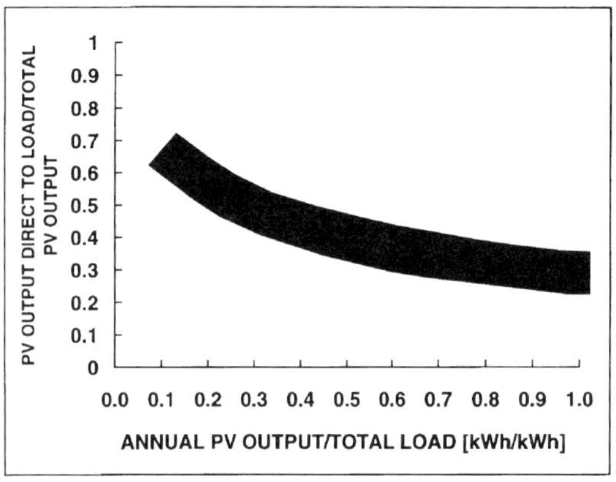


*Figure 17.2 Average fraction of directly used PV energy from the total house load as a function of the annual PV electricity output/annual load ratio in Europe and North America with a typical one-family house load profile.*

In grid-connected systems this efficiency mainly depends on inverter and wiring losses. Typically the wiring losses are 10% and the inverter losses 15%.

Thus the $\eta_{BOS}$ is approximately 75%. The annual radiation S can be taken from Worksheet #1.

When the total energy production $E_{PV}$ has been determined, the ratio of $E_{PV}$ and $E_{load}$ can be calculated. With this value and Figure 17.2, the amount of directly used energy is estimated.

It must be noted that Figure 17.2 is an average result obtained with average house load profiles and insolation data for Europe and North America. The calculations can also be performed vice versa by starting from a wanted ratio of directly used PV energy back to PV array size.

[1] Note: In order to achieve correct results, mentioned units must be used in all formulae and worksheets.

| Location | Inverter power ratio $P(DC)_{Inverter}/P_{PV}$ |
|---|---|
| Northern Europe (55-70°N) | 0.7 - 0.8 |
| Central Europe (45-55°N) | 0.75 - 0.9 |
| Southern Europe (35-45 °N) | 0.85 - 1.0 |

*Table 17.1 Recommended inverter sizes for different locations.*

The nominal power of the inverter should be smaller than the PV nominal power. The optimum ratio depends on the climate, the inverter efficiency curve and the inverter/PV price ratio. Computer simulation studies indicate a ratio $P\ (DC)_{Inverter}/P_{PV}$ of 0.7-1.0. The recommended inverter sizes for different locations are shown in Table 17.1.

In order to list metering options and other basic information on the grid-connected case as well as for performing the sizing calculations, Worksheets #3 part I and II are prepared and may be found in Appendix II.

**Sizing of a stand-alone PV-battery system (Worksheets #4 & 5 & 6)**

For stand-alone PV battery systems the sizing must be more accurate than for grid-connected systems, because the available buffer capacity is quite limited. To compensate for unexpected long cloudy periods some oversizing of the battery size as well as of the PV array size is needed. This oversizing also reduces the average battery 'Depth of Discharge' (DOD) and thus increases the battery life.

After performing the load estimation with the help of Worksheet #2, the required autonomy time is chosen. The autonomy time varies from case to case and depends on latitude, operation season and required percentage of availability (safety margin). In Worksheet #4 (Appendix II) recommendations of autonomy time for different locations are given. Battery capacity is also dependent on discharge current and temperature. In Worksheet #4 this factor is shown as a function of the discharge rate and average storage temperature of the month in which storage is needed most. This derating is especially important, if the battery is located outdoors in cold climates. The maximum allowable DOD depends on battery type and load profile. For a typical lead-acid battery this fraction is between 0.5-0.8.

The next step is to size the PV array and the other system components. This is done with the help of Worksheet #5. For PV array sizing the month with the lowest insolation on the array plane is chosen as the design month (from Worksheet #1).

Dividing the average daily load of the design month by the average daily solar insolation and the system component efficiencies, yields the necessary PV array size (kW). The efficiencies to be taken into account are wiring efficiencies (typically 90%), charge regulator efficiency (typically 85%) and battery efficiency (typically 90%). A safety margin is recommended and presented in Worksheet #5. The design array current and the size of the power conditioning unit is then estimated.

The PV array area corresponding to the calculated array power is estimated as in the grid-connected case. The module efficiency is module-type dependent. For rough calculations, average module efficiencies as presented in Table 17.2 can be used.

At this stage it must be confirmed whether a PV battery system is enough to satisfy the load or whether a back-up generator is needed. This can be done with Worksheet #6 'Consider Hybrid', where the array to load ratio is calculated and used for this decision.

| Module type | Average efficiency |
|---|---|
| Amorphous | 5 % |
| Mono-crystalline | 13 % |
| Polycrystalline | 12 % |

*Table 17.2 Average efficiencies of different commercial silicon-based PV module types.*

**Sizing of a stand-alone PV hybrid system (Worksheet #7)**

For a PV-hybrid system the PV array sizing procedure is similar to that used for a PV-battery system, but now the battery - generator pair must be sized so that it can back-up the shortfall during the month with the lowest insolation (wintertime).

An experienced designer should be consulted to decide whether a back-up generator is needed. Generally, if there are large seasonal variations in available solar radiation or long overcast periods or if there is a need to supply the load all the time, the back-up generator will be specified.

Figure 17.3 shows a simple chart as guideline for this decision. The figure is based on practical experiences with existing systems. According to the figure, a hybrid system should be preferred when the load is large. This is mainly because of cost considerations. Also, if the PV array size obtained for a corresponding stand-alone system is large compared to the load, a hybrid system will be most economical and practical.
Examples of sizing procedures for different types of PV systems are presented in sections 17.4 - 17.6.

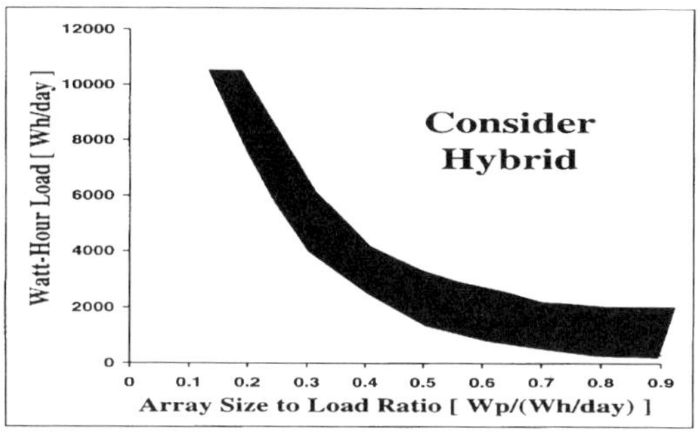


*Figure 17.3 A graph showing the relationship of array to load ratio and the load size when a hybrid system should be used. (Courtesy of CANMET/NRCan.)*

## 17.3 Design and simulation programs

In addition to the simple sizing tools, such as worksheets previously presented, several computer programs have been developed to design or simulate PV systems.

A computer design software will, like a worksheet, use statistical information about the environmental conditions where a system is set and its load, to determine the optimum size of its components. This information is usually provided on a monthly basis. The program then uses a fixed control algorithm and a specific system configuration to calculate the size of the system and its components. Most design software will also calculate the cost of the system, provided that it includes detailed information on the cost of each component.

A computer simulation program will usually be used to calculate the performance of a given system under variable operational conditions. These operational conditions are usually given in short constant time steps (usually hours or days). The program uses reasonable accurate empirical mathematical models of different system components to simulate the actual operation of the system for a fixed period of time.
By changing the input data and control algorithms, the user can then optimize the perfor-

mance of his system. He can also modify the size and nature of any individual component to achieve the best performance under the simulated operational conditions.

While simulation programs are most suitable for research purposes to perform sensitivity analysis and to study different control strategies, they are very difficult to use when designing systems. This is because they require inputs, which the designer does not readily have and are usually highly variable. Also an experienced designer and user of the programs is usually needed to interpret the results correctly.

There exists a number of softwares available from different companies. Some of the simulation programs and design tools are proprietary for companies, but there exist also commonly available tools. Table 17.3 lists an overview of PV simulation and design programs.

| **Program** | **Type** | **PV** | **Wind** | **Stand-alone** | **Grid-connected** | **Facade** | **Pump** |
|---|---|---|---|---|---|---|---|
| ASHLING | SIM | + | - | + | + | - | - |
| DAST-PVPS | SIM | + | - | + | - | - | + |
| FWISO 81 | SIM | + | + | + | + | - | + |
| INSEL | SYS | + | + | + | + | - | + |
| ISEE | DB | + | + | - | - | - | - |
| ITE-BOSS | SYS | + | - | + | + | - | + |
| PHOTO | SIM | + | + | + | + | - | - |
| PV 1.03 | SIM | + | - | o | + | - | - |
| PVcalc 1.03 | SIM | + | - | o | + | - | - |
| PV-DIMM | SIM | + | - | + | - | - | - |
| PV F-CHART | TAB | + | - | + | + | - | - |
| PV-TAS | SIM | + | - | + | + | + | - |
| PVDIM | SIM | + | - | + | + | - | + |
| PVFORM | SIM | + | - | + | + | - | - |
| PVnode | SIM | + | - | + | + | + | - |
| PVPUMP | SIM | + | - | + | - | - | + |
| PVS | SIM | + | - | + | + | - | - |
| PVSHAD | SIM | + | - | + | + | + | - |
| SHADE | SIM | + | - | + | + | + | - |
| SOMES | SIM | + | + | + | + | - | - |
| SYSTEMSPEC | SIM | + | - | + | + | - | - |
| TRNSYS | SYS | + | + | + | + | + | - |
| WAsP | TAB | - | + | + | + | - | - |
| WATSUN-PV | SIM | + | - | + | + | - | + |

+ yes, o conditional, - no

Type of program:

| | |
|---|---|
| TAB | Statistics-based programs |
| SIM | Time step simulators |
| SYS | Simulation systems |
| DB | Databases |

Type of system:

| | |
|---|---|
| PV | PV generators |
| Wind | Wind generators |
| Stand-alone | Stand-alone systems |
| Grid-connected | Grid-connected systems |
| Facade | Inhomogeneously irradiated generator surfaces (facades) |
| Pump | PV pump systems |

*Table 17.3 Overview of PV simulation programs.*

## 17.4 Sizing example of a grid connected PV building system

The general description of a grid-connected example system is as follows:

| | |
|---|---|
| Location: | Stuttgart, 49°N, 9°E, Germany |
| Shadow: | 10° horizon shadowing |
| House: | typical one-family house, the roof faces south with a tilt angle of 45°, the available roof area is 40 $m^2$ |
| Load: | 3000 kWh/year AC electricity, (energy-efficient house). |

First, the solar availability on the roof surface must be defined with the help of Worksheet #1. From Appendix I the solar radiation data of Stuttgart can be found. For monthly tilt, azimuth and shadow angle factors the data of Madison will be used as the best approximate. The 10° shadow of the horizon can be taken into account by calculating the average of the two shadow angle figures for 0° and 20°.

Finally, by simple multiplications, the monthly figures can be written into the Worksheet. Summing up the monthly figures gives a radiation value of 1202 kWh/$m^2$ on the PV roof surface.

In this case it is not necessary to do any load analysis and efficiency improvements because the house is fairly new and efficient already and previous electricity bills have shown that the electricity consumption is 3000 kWh/year and the profile can be assumed to be roughly constant in this case.

Worksheet #3 is then used to size the system components. Let us choose as the PV array nominal power a typical 2 kW size. The PV array area with monocrystalline modules is approximately 16 $m^2$. A typical average value for a grid-connected inverter efficiency is 85%. Wiring losses are usually 10%. Thus the annually produced energy is 1655 kWh. The directly used PV energy fraction is 30-50%. This means that roughly 60% of the PV electricity is sold to the grid annually. The optimum inverter size is 1.5-1.8 kW. In practice, the available inverters are limited to certain sizes and if a suitable inverter is not found the PV array size might be adjusted slightly.

GRID-CONNECTED EXAMPLE CASE

**WORKSHEET #1: DEFINE SITE CONDITIONS AND SOLAR AVAILABILITY**

| | | | |
|---|---|---|---|
| SYSTEM: "Stuttgart" PV house | | | |
| SYSTEM LOCATION: | LATITUDE: 48.8 °N | LONGITUDE: 9 °E | |
| INSOLATION LOCATION: | LATITUDE: 48.8°N | LONGITUDE: 9 °E | |

| MONTH | Location: Ambient temperature | Location: Horizontal insolation | * | Array plane: tilt, azimuth, shadow Factor (appendix I) | = | Array plane: Insolation | * | = | Array plane: Insolation |
|---|---|---|---|---|---|---|---|---|---|
| | °C | kWh/m²day | * | fraction | = | kWh/m²day | * | = | kWh/m²month |
| January | 0 | 1.0 | * | 1.5 | = | 1.5 | *31 | = | 46.5 |
| February | | 1.7 | * | 1.35 | = | 2.295 | *28 | = | 64.26 |
| March | 10 | 2.7 | * | 1.2 | = | 3.24 | *31 | = | 100.44 |
| April | | 4.1 | * | 1.0 | = | 4.1 | *30 | = | 123 |
| May | | 5.0 | * | 0.9 | = | 4.5 | *31 | = | 139.5 |
| June | 19 | 5.4 | * | 0.9 | = | 4.86 | *30 | = | 145.8 |
| July | | 5.4 | * | 0.9 | = | 4.86 | *31 | = | 150.66 |
| August | | 4.5 | * | 1.0 | = | 4.5 | *31 | = | 139.5 |
| September | 10 | 3.6 | * | 1.1 | = | 3.96 | *30 | = | 118.8 |
| October | | 2.2 | * | 1.3 | = | 2.86 | *31 | = | 88.66 |
| November | | 1.1 | * | 1.4 | = | 1.54 | *30 | = | 46.2 |
| December | 0 | 0.8 | * | 1.55 | = | 1.24 | *31 | = | 38.44 |
| S = Annual insolation on PV array [kWh/m²] = Σ | | | | | | | | = | 1202 |

GRID-CONNECTED EXAMPLE CASE

**WORKSHEET #3: GRID CONNECTED SYSTEM (part I)**

| Chosen PV array power $P_{PV}$ [$kW_p$] | / | PV efficiency (table 17.2) $\eta_{PV}$ [fraction] | = | PV array area $A_{PV}$ [m²] |
|---|---|---|---|---|
| 2.0 | / | 0.13 | = | 15.38 |

| Chosen PV Array power $P_{PV}$ [$kW_p$] | * | Annual insolation on PV array (worksheet #1) S [kWh/m²] | * | BOS efficiency (see below) $\eta_{BOS}$ [fraction] | * | $K_{PV}$ factor [fraction] | = | Annual produced PV energy $E_{PV}$ [kWh] |
|---|---|---|---|---|---|---|---|---|
| 2.0 | * | 1202 | * | 0.765 | * | 0.9 | = | 1655 |

| Annual produced PV energy $E_{PV}$ [kWh] | / | Annual load energy (worksheet #2) [kWh] | = | PV/load ratio [fraction] | from Figure 17.3 | Directly used PV energy [fraction] |
|---|---|---|---|---|---|---|
| 1655 | / | 3000 | = | 0.55 | ==> | 0.3 - 0.5 |

| Chosen PV Array power $P_{PV}$ [$kW_p$] | * | Optimum inverter size (from table 17.1) [fraction] | = | Inverter nominal power [kW] |
|---|---|---|---|---|
| 2.0 | * | 0.75 - 0.9 | = | 1.5 - 1.8 |

| average inverter efficiency [fraction] | * | wiring loss factor (1-loss fraction) [fraction] | = | BOS efficiency $\eta_{BOS}$ [fraction] |
|---|---|---|---|---|
| 0.85 | * | (1 - 0.1) | = | 0.765 |

*Figure 17.4 Filled-in Worksheets #1 and #3 (part I) for the grid-connected PV system example described in Section 17.4.*

## 17.5 Sizing example of an autonomous PV building system

The following case shows an example of an autonomous PV system used to power a remote vacation cabin. The cabin is situated in the heart of the Laurentians near Montreal, Canada, and it is used mainly on weekends and holidays during the summer as well as on an occasional weekend in winter. The cabin is one kilometre away from the grid and the owner enjoys the fact that his cabin is totally isolated. Still, after years of using oil lamps and hauling water by hand, the owner would like to enjoy the benefits of electricity. The description of the case is as follows:

Location: Bark Lake; 46°N, 74°W, Canada

Shadow: 20° horizon shadowing

House: Small cabin with roof face tilted at 45° angle due south, overlooking a lake in front. Some trees are partially shading the array in the morning and afternoon in the summer

Loads: Variable, average of 980Wh/day when inhabited. The owner wants to use ordinary AC appliances except for his lighting and refrigeration needs. In winter, the cabin is heated with wood.

The solar availability is estimated using Worksheet #1 with the solar radiation data (Montreal) and the correction factors (Madison) found in Appendix I.

AUTONOMOUS EXAMPLE CASE

**WORKSHEET #1: DEFINE SITE CONDITIONS AND SOLAR AVAILABILITY**

SYSTEM: Bark Lake house

| SYSTEM LOCATION: | LATITUDE: 46 °N | LONGITUDE: 74 °W |
|---|---|---|
| INSOLATION LOCATION: | LATITUDE: 45.5 °N | LONGITUDE: |

| MONTH | Location | | | Array plane | | | | | |
|---|---|---|---|---|---|---|---|---|---|
| | Ambient temperature | Horizontal insolation | * | tilt, azimuth, shadow Factor (appendix I) | = | Insolation | | | |
| | °C | kWh/m²day | * | fraction | = | kWh/m²day | * | = | kWh/m²month |
| January | -10 | 1.5 | * | 1.4 | = | 2.1 | *31 | = | 65.1 |
| February | | 2.4 | * | 1.3 | = | 3.12 | *28 | = | 87.4 |
| March | | 3.5 | * | 1.2 | = | 4.2 | *31 | = | 130.2 |
| April | 6 | 4.4 | * | 1.0 | = | 4.4 | *30 | = | 132 |
| May | | 5.3 | * | 0.9 | = | 4.77 | *31 | = | 147.9 |
| June | 21 | 5.6 | * | 0.9 | = | 5.04 | *30 | = | 151.2 |
| July | 21 | 5.8 | * | 0.9 | = | 5.22 | *31 | = | 161.8 |
| August | | 4.8 | * | 1.0 | = | 4.8 | *31 | = | 148.8 |
| September | 6 | 3.7 | * | 1.1 | = | 4.07 | *30 | = | 122.1 |
| October | | 2.2 | * | 1.3 | = | 2.86 | *31 | = | 88.66 |
| November | | 1.3 | * | 1.3 | = | 1.69 | *30 | = | 50.7 |
| December | -10 | 1.1 | * | 1.5 | = | 1.65 | *31 | = | 51.15 |

S = Annual insolation on PV array [kWh/m²] = Σ = 1337.01

*Figure 17.5 Filled-in Worksheet #1 for the autonomous PV systems example described in Section 17.5.*

AUTONOMOUS EXAMPLE CASE

**WORKSHEET #2: ESTIMATE LOADS**

| Load Description | AC or DC | AC loads (1) [W] | Inverter efficiency (2) [%] | DC load (3)=(1)/(2) [W] | Duty cycle (4) [h/day] | Duty cycle (5) [day/week] | Daily load (6)= (3)*(4)*(5)/7 [Wh/day] | Nominal voltage (7) [V] | Ah-Load (8)=(6)/(7) [Ah/day] |
|---|---|---|---|---|---|---|---|---|---|
| Bedr. lights (2) | DC | | | 26 | 2 | 2 | 14.86 | 24 | 0.62 |
| Livingr. lights (3) | DC | | | 39 | 5 | 2 | 55.71 | 24 | 2.32 |
| Kitchen lights (2) | DC | | | 26 | 2 | 2 | 14.86 | 24 | 0.62 |
| Microwave oven | AC | 375 | 0.9 | 417 | 1 | 2 | 119.05 | 24 | 4.96 |
| Water pump | DC | | | 250 | 0.5 | 2 | 35.71 | 24 | 1.49 |
| Refrigerator | DC | | | 250 | 2 | 7 | 500.00 | 24 | 20.83 |
| Stereo | AC | 30 | 0.8 | 38 | 1 | 2 | 10.71 | 24 | 0.45 |
| Color TV | AC | 150 | 0.9 | 167 | 2 | 2 | 95.24 | 24 | 3.97 |
| Miscellan. loads | AC | 1000 | 0.85 | 1176 | 0.4 | 2 | 134.45 | 24 | 5.60 |
| | | | | | | | | | |
| | | MAXIMUM DC LOAD (9) [W] | | 2388 | TOTAL DAILY LOADS (10)=Σ(6) [Wh/day] | | 980.6 | TOTAL LOAD (11)=Σ(8) [Ah/day] | 40.86 |

| | | |
|---|---|---|
| DESIGN LOAD (Total load=(11)) | 41 | Ah/Day |
| DESIGN PEAK CURRENT DRAW (Maximum DC load / Nominal Voltage) | 100 | A |
| ANNUAL LOAD ENERGY (Total daily load *weeks in operation*7/1000 | 268 | kWh |

NOTE: Design calculations are made for Spring, Summer and Fall seasons. In winter, the system is seldom used and it is assumed that the system will recuperate fully between each usage

AUTONOMOUS EXAMPLE CASE

**WORKSHEET #4: SIZE BATTERY BANK**

| Design load (worksheet #2) [Ah/day] | * | Days of autonomy (see appendix 17.1) [Days] | / | Max depth of discharge [fraction] | = | Usable battery capacity [Ah] |
|---|---|---|---|---|---|---|
| 41 | * | 2 | / | 0.3 | = | 273 |

| | | |
|---|---|---|
| OPERATING TEMP = | 15 | [degrees C] |
| DISCHARGE RATE = | 24 x DAYS OF AUTONOMY | |
| = | 48 | [h] |

| Usable battery capacity [Ah] | / | Usable fraction of capacity available (from graph below) | = | Design battery capacity [Ah] |
|---|---|---|---|---|
| 273 | / | 0.95 | = | 287 |

*Figure 17.6 Filled-in Worksheets #2 and #4 for the autonomous PV systems example described in Section 17.5.*

AUTONOMOUS EXAMPLE CASE

**WORKSHEET #5: SIZE ARRAY & COMPONENTS**

| OPERATING SEASON (Months) | February - October |
|---|---|

| Design month daily load [kWh/day] | / | Lowest insolation on PV array (worksheet #1) [(kWh/m²)/day] | / | Wiring loss factor [fraction] | / | Charge regulator efficiency [fraction] | / | Battery efficiency [fraction] | = | Design PV array power [kWp] |
|---|---|---|---|---|---|---|---|---|---|---|
| 0.98 | / | 2.86 | / | 0.9 | / | 0.85 | / | 0.9 | = | 0.498 |

| Design PV array power [Wp] | * | PV array sizing safety factor (see table below) | = | PV array power [Wp] |
|---|---|---|---|---|
| 498 | * | 1.3 | = | 647 |

| PV array power [Wp] | / | Nominal voltage [V] | = | Design array current [A] |
|---|---|---|---|---|
| 647 | / | 24 | = | 27 |

| Design array current [A] | = | Design power conditioner current [A]* |
|---|---|---|
| 27 | ≠ | 100 |

peak load current is higher than Dedign array current

| PV array power $P_{PV}$ [kWp] | / | PV module efficiency (from table 17.2) $\eta_{PV}$ [fraction] | = | PV array area $A_{PV}$ [m²] |
|---|---|---|---|---|
| 0.647 | / | 0.12 | = | 5.39 |

*Figure 17.7 Filled-in Worksheet #5 for the autonomous PV system example described in Section 17.5*

The next step is to define the load for the cabin. Using the load analysis portion of Worksheet #2 the average total load per day is found to be around 980 Wh/day. Note that the total load per day can be as high as 2100 Wh/day during the weekend. The designer should then choose deep cycle batteries because the batteries may discharge completely during the weekend.

Since the owner wants AC loads and the more efficient inverters are working at 24 V, the system will be a 24 V system. Note that the refrigerator is a large part of the load and that it is turned on throughout the summer. However, it is turned off during winter and only put on when the owner is using the cabin.

Since the owner is planning to use his cabin only on weekend holidays, most of the load will be at these times. This in turn means that the battery bank can be relatively small since it would normally be recharged during the week. But, for the same reason, the designer should use a low average for the Maximum Depth of Discharge (DOD) when estimating the capacity of the battery. We can thus calculate an average maximum depth of discharge for the battery at 30% and two days of autonomy. Note that the battery will be located inside a special shed but since the cabin will not be heated in winter, the battery should be allowed to recharge throughout the period. Since little consumption will be drawn from the battery, the system should be well protected against overcharge. The battery sizing is presented in Worksheet #4.

The final step of the sizing (Worksheet #5) requires calculation of the lowest insolation month with the corresponding highest load month. In our example, it is obvious that March or September will be the most demanding months for the system and the array should thus be calculated for these months. In this application the efficiency of the battery must be taken into account.

Thus the resulting system design is as follows:

| | |
|---|---|
| PV array size: | 650 Wp |
| PV array area: | 5.4 $m^2$ |
| Design array current: | 27 A |
| Inverter size: | 1600 VA |
| Battery size: | 300 Ah |

## 17.6 Sizing example of a hybrid PV building system

As an example case of a hybrid PV building system a house located near the Helsinki coast on a small island in Southern Finland will be used. The house is owned by a fisherman and his family. The house is inhabited all year round. The description of the case is as follows:

| | |
|---|---|
| Location: | Island near coast, 60°N, 23°E, Finland |
| Shadow: | 0° south, 15° north |
| House: | typical one-family house, the roof faces south-east with a tilt angle of 30°, the available roof area is 60 $m^2$ |
| Load: | roughly constant load profile, 2500 kWh/year AC electricity. |

Different approaches can be used to size a photovoltaic-diesel hybrid system. One approach is to size the system assuming that photovoltaics will provide a given percentage of the system electricity need.

Here it is assumed that photovoltaics will satisfy the main electricity demand during certain months, namely from April to September.

HYBRID EXAMPLE CASE

**WORKSHEET #1: DEFINE SITE CONDITIONS AND SOLAR AVAILABILITY**

| | | | |
|---|---|---|---|
| SYSTEM: Island system near Helsinki | | | |
| SYSTEM LOCATION: | LATITUDE: 60 °N | LONGITUDE: | |
| INSOLATION LOCATION: | LATITUDE: 60 °N | LONGITUDE: | |

| MONTH | Location | | | Array plane | | | | | |
|---|---|---|---|---|---|---|---|---|---|
| | Ambient temperature | Horizontal insolation | * | tilt, azimuth, shadow Factor (appendix I) | = | Insolation | | | |
| | °C | kWh/m²day | * | fraction | = | kWh/m²day | * | = | kWh/m²month |
| January | -10 | 0.3 | * | 1.4 | = | 0.42 | *31 | = | 13.02 |
| February | | 0.8 | * | 1.3 | = | 1.04 | *28 | = | 29.12 |
| March | | 2.2 | * | 1.1 | = | 2.42 | *31 | = | 75.02 |
| April | 4 | 3.4 | * | 1.1 | = | 3.74 | *30 | = | 112.2 |
| May | | 5.1 | * | 1.0 | = | 5.1 | *31 | = | 158.1 |
| June | 16 | 6.0 | * | 1.0 | = | 6.0 | *30 | = | 180 |
| July | 16 | 5.2 | * | 1.0 | = | 5.2 | *31 | = | 161.2 |
| August | | 4.1 | * | 1.1 | = | 4.51 | *31 | = | 139.81 |
| September | | 2.3 | * | 1.1 | = | 2.53 | *30 | = | 75.9 |
| October | 4 | 1.0 | * | 1.3 | = | 1.3 | *31 | = | 40.3 |
| November | | 0.3 | * | 1.4 | = | 0.42 | *30 | = | 12.6 |
| December | -10 | 0.2 | * | 1.7 | = | 0.34 | *31 | = | 10.54 |

S = Annual insolation on PV array [kWh/m²] = Σ = 1008

*Figure 17.8 Filled-in Worksheet #1 for the hybrid PV system example described in Section 17.6.*

HYBRID EXAMPLE CASE

WORKSHEET #7: SIZE HYBRID

| | | |
|---|---|---|
| PV DESIGN PERIOD | April - September | Months |
| LOWEST INSOLATION DURING PV DESIGN PERIOD (from worksheet #1) | | $kWh/m^2day$ |

| Design month daily load (worksheet #2) [kWh/day] | / | Lowest insolation on PV array during PV design period (worksheet #1) [$(kWh/m^2)$/day] | / | Wiring loss factor [fraction] | / | Charge regulator efficiency [fraction] | / | Battery efficiency [fraction] | = | PV array power [$kW_p$] |
|---|---|---|---|---|---|---|---|---|---|---|
| 6.85 | / | 2.53 | / | 0.9 | / | 0.85 | / | 0.9 | = | 3.93 |

| PV array power [$W_p$] | / | Nominal voltage [V] | = | Design array current [A] |
|---|---|---|---|---|
| 3930 | / | 48 | = | 82 |

| Design array current [A] | * | Lowest insolation on PV array (worksheet #1) [$(kWh/m^2)$/day] | = | PV load contribution [Ah/day] |
|---|---|---|---|---|
| 82 | * | 0.34 | = | 28 |

| Design load (worksheet #2) [Ah/day] | * | Days of autonomy [days] | / | Maximum depth of discharge [fraction] | / | Usable fraction of* battery capacity available (from worksheet #4) | = | Design battery capacity [Ah] |
|---|---|---|---|---|---|---|---|---|
| 142 | * | 2 | / | 0.5 | / | 0.95 | = | 598 |

* 2 days autonomy gives approximately 50 h discharge rate, batteries located at 15 °C during PV design period

| Design load (worksheet #2) [Ah/day] | / | Battery efficiency [fraction] | / | Rectifier efficiency [fraction] | * | Nominal voltage [V] | = | Design generator load [Wh/day] |
|---|---|---|---|---|---|---|---|---|
| 142 | / | 0.9 | / | 0.8 | * | 48 | = | 9467 |

| Design generator load [Wh/day] | * | Days of autonomy (see above) [days] | / | Charge time [h] | = | Nominal generator capacity [W] |
|---|---|---|---|---|---|---|
| 9467 | * | 1 | / | 4 | = | 2367 |

*Figure 17.9 Filled-in Worksheet #7 for the hybrid PV system example described in Section 17.6.*

The first steps for this case are the same as in the previous cases, starting from solar availability estimations with Worksheet #1 and the solar radiation data and correction factors of Appendix I. For correction factors the data for Copenhagen will be used as the best approximate.

The next step would be to use Worksheet #2 for the load estimation. However, the load is given here to be constant 2500 kWh/year, which means 6.85 kWh/day. This results in 6850/48 Ah/day = 142 Ah/day with system voltage of 48 V. The following steps would be to use Worksheets #4 and #5 to estimate the battery and the PV array size and with Worksheet #6 whether a PV-hybrid system should be considered. These steps are omitted here, because in this very northern example (latitude 60°N) it is evident that a diesel generator is needed if the load has to be guaranteed for the whole year. This can be read from Worksheet #1: during the three darkest winter months the available solar radiation is very small. Thus the sizing can be finished with Worksheet #7. The batteries and diesel genset are located in an insulated cabin so that the battery temperature during the PV design period is roughly 15°C.

The sizing results are:

| | |
|---|---|
| PV array size: | 4000 Wp |
| Design array current: | 82 A |
| Battery size: | 610 Ah |
| Diesel generator size: | 2400 kW |

The PV array area with polycrystalline cells is estimated to be 4.0 $kW_p$/0.12 = 33 $m^2$.

# Chapter 18

# Key Component Selection

## 18.1 Introduction

Once the PV system has been sized as explained in Chapters 15 through 17, the installation may be planned. This means designing the physical layout of the system, selecting the proper equipment to meet the design requirements and ordering the different parts. In this chapter, criteria and guidelines for the design of a proper layout and selection of equipment available on the market are provided.

The designer should specify components in the following order:

1. Choose place and mounting method for modules, select modules;
2. Choose place for batteries, select battery type (off-grid systems only);
3. Select the necessary power conditioning equipment and inverter;
4. Select back-up system (if needed);
5. Estimate overall system losses and respecify components, if necessary;
6. Specify safety devices and switchgears;
7. Do layout of wiring, specify size and type;
8. Prepare a full list of parts and tools to order.

## 18.2 Selection of the photovoltaic modules

Photovoltaic modules come in different types, sizes and shapes. During the sizing procedure presented in Chapter 17, the array size has been determined in terms of peak watts delivered at peak sun hours. The designer must now select the actual photovoltaic module type to be used and calculate the number of modules in the array. Physical considerations such as available area, mounting structure type and architectural aspects limit the size of the array and influence the selection of the type of module to be used. Inclination, shadowing and ventilation of the modules will affect the electrical characteristics of the array and may change the design size of the system.

**General position of the modules**

Photovoltaic modules should always be placed close to the control unit and batteries to minimize voltage drop along wires. Low voltage with high direct currents lead to high ohmic losses and require large size wires. These are bulky, very expensive and hard to work with.

The modules should be mounted in a place where there is no or only a minimum of shading from surrounding objects such as trees and other buildings. Even a small shadow on just one cell can affect the performance of the whole array. If the array cannot be placed without being affected by shading, sizing the array will have to take into account the loss of electricity production due to this shading effect. This in turn will affect the quantity of modules to be used.

**In general, if partial shading cannot totally be avoided, the series strings should be arranged in such a way that only one of them is affected.**

The array should be positioned in a direction and inclination to produce maximum power during the time when it is needed most and when power production is maximum. These times may not coincide and during the sizing process adjustments must be made to size the system for its optimal design.

**Mounting of the array**

Photovoltaic modules generate most electricity when facing the sun directly, but the position of the sun changes through the day. It is possible to mount the modules on trackers to improve the power output of the array by allowing the tilt angle and direction of the modules to follow the sun. Their use may significantly improve the power output of the array in regions where direct sunlight is prevalent and thus reduce the size of the array. However, trackers have almost no impact in regions where diffuse sunlight accounts for more than 50% of the total insolation. Trackers can be very expensive and they are difficult to integrate in a built environment.

Integration of PV modules in a building surface may influence the size of the modules used, the size of the array, its inclination and its direction. When sub-arrays have different sizes and mounting directions, special care must be given to optimize power production so that each sub-array can deliver maximum power to the load.

Other mounting considerations which may affect the size and location of the array are:

- Photovoltaic arrays must be strong enough to withstand wind loads and snow accumulation.
- The modules must be held securely for their lifetime which is estimated to be 20-30 years.
- Ease of access to the modules for maintenance and cleaning must be planned at the design stage.

**Module types**

Many different types of PV modules are commercially available. More efficient modules will lead to smaller arrays.

- Modules made of crystalline silicon cells are most widely used; their efficiencies range from 12 to 15%.
- Modules made of thin-film amorphous silicon are cheaper to produce but their efficiencies range from 5 to 10% only.
- Thin-film materials such as cadmium telluride (CdTe) and copper indium diselenide (CIS) are not yet fully commercialized in modules and still expensive although they promise to become a low cost solution. Their efficiencies range from 7 to 10%.

**Architectural aspects**

PV modules can have different shapes and appearances depending on the material they are made of and the way they are produced. Modules made of crystalline cells are made of round or square cells encapsulated in glass with a clear or coloured back side. Amorphous silicon cells are usually dark red. Thin-film modules can be micro-perforated to give a defined transmissivity, but this will reduce their efficiencies.

**Frame**

PV modules come with different frame types and colours and can also be produced frameless to facilitate their integration into a building.

**Module size**
Choosing the largest module size available will reduce the amount of handling and installation time. There will be fewer electrical and mechanical joints which should improve the tightness of building mounted arrays. On the other hand, large modules are heavy and difficult to handle and can be expensive to replace in case of an accident.

**Modules in series**
The number of modules in series is determined by dividing the designed system voltage (usually determined by the battery bank or the inverter) with the nominal module voltage that occurs at lower temperatures. The nominal module voltage is often a multiple of 12 V.

Since module voltage decreases with increasing temperature (in the order of 2.2 mV/K/ cell for Si), care must be given when integrating the module to a building so that the modules do not overheat because of lack of ventilation. The maximum power point of the series string must be calculated to be in the range of the design system voltage for all operating temperatures.

The number of modules in series, $n_s$, is calculated by

$$n_s = \frac{V_{system,max.}}{V_{module,corr.}}$$

where

$V_{system,max.}$ is the maximum system voltage (at maximum charging current) and
$V_{module,corr.}$ is the module voltage corrected for operating conditions.

**Strings in parallel**
The number of modules in parallel is determined by dividing the designed array output ($W_p$) by the selected module output ($W_p$) and the number of strings.

$$n_p = \frac{P_{array,max.}}{n_s * P_{module,max.}}$$

where

$P_{array,max.}$ is the peak array power and
$P_{module,max.}$ is the module peak power.

Both, the number of modules per series string and the number of parallel strings must be integers. Once the numbers of modules in parallel and in series are known, the total number of modules in the array is found by their product.

## 18.3 Selection of the storage component (off-grid systems only)

During the design process, it must be decided whether a storage component has to be used in the proposed system. Stand-alone building PV systems always need some form of storage. Notwithstanding the use of mechanical devices to store energy (such as elevating water or compressing air), electrochemical secondary batteries are presently the only commercial means to store electricity for a period of time exceeding a few days. The required battery capacity and nominal battery (system) voltage has been estimated during the sizing process. The decision to use deep or shallow cycling batteries must be made according to the load requirement.

The selection of the type of battery depends on the operation of the system and the environment of the battery. It is important to choose a quality battery well suited for the application, since the battery is the key element which defines the lifetime of the system and most of the maintenance requirements. Care must be taken not to mix batteries with different characteristics, since this will negatively affect the overall performance of the whole system. For the same reason, new batteries should not be used in combination with old ones.

It is important to keep batteries at a moderate temperature of 10-25°C for optimum performance. In hot climates, the electrolyte in a battery has a tendency to evaporate or cause deformation of the battery and its efficiency is reduced. Electrode corrosion is also accelerated. In cold climates, the battery is more difficult to charge, the electrolyte may freeze and the battery can be destroyed. Batteries will also need a higher voltage to complete charging in cold weather. Properly insulated enclosures help keep temperature swings from affecting the battery. However, this enclosure must be well ventilated, especially in the case of liquid electrolytes, since charging the battery causes some hydrogen production and may cause a deflagration if it is allowed to accumulate. If ventilation is not possible, the use of either gel-type sealed batteries or glass fibre matted batteries will permit the use of an enclosure to protect the batteries.

Protection against vandalism and harmful damage are other considerations which affect the type of battery and its enclosure. Batteries are expensive items and could be used for other than the intended purposes. The use of a battery bank made of 2 V cells will deter thieves looking for suitable vehicle batteries. Large battery capacities will also benefit from using individual 2 V cells which are easier to handle.

Chapter 7 has shown the characteristics of different battery types. In general:

- Lead-acid batteries should be selected for large battery banks because of the lower initial costs. Lead-acid batteries are the most common and the least expensive type of battery. Note that lead-acid batteries are nominally rated at 2 V per cell but that their operation cycle lies between 1.75 to 2.45 V. Equipment connected to these batteries must accept this range.

- Lead-acid batteries with calcium alloy should be selected for shallow cycling with long autonomy periods (50 to 500 hours discharge time) since they are virtually maintenance free.

- Lead-acid batteries with low antimony alloy (0.1 - 2% antimony) should be selected for deep-cycling and daily discharging.

- Lead-acid batteries should have a higher concentrated electrolyte when used in a cold environment (specific gravity of 1.3 $kg/dm^3$ instead of 1.25 $kg/dm^3$). This influences the charging voltage but keeps the battery from freezing when deeply discharged.

- Nickel-cadmium batteries are less affected by extreme temperatures and supply a constant voltage through most of their discharge cycle. They also withstand a higher number of cycles than lead-acid batteries. Note that a Ni-Cd cell is rated at 1.2 V and thus requires more cells than a lead-acid battery for the same voltage.

- Gel-type or AGM (Absorptive Glass Mat) batteries should be used in areas where transportation is difficult and where ventilation of the batteries is a problem. Note that these batteries can be positioned in any way and are easier to stack.

After selection of the battery type has been completed, the capacity requirements should be re-evaluated to include the efficiency and electrical characteristics of the battery selected. In choosing the size of the battery, select the cell with the amp-hour (Ah) rating nearest to the one calculated. Then determine the number of cells to be connected in series by dividing the nominal system voltage by the nominal cell voltage. Consider that it may be cheaper to choose a battery cell with half the capacity and use two parallel strings of batteries. Larger capacity cells can be more expensive per capacity and are bulky, heavy and difficult to handle.

## 18.4 Selection of the power conditioning equipment

Proper selection of the power conditioning equipment ensures that the system will operate in its optimum range and also extends the lifetime of the various components. Improper selections can result in inefficiency, faulty operation, safety hazard, inadequate performance and excessive costs. Included in the general term 'Power conditioning equipment' are battery charge regulators, load controllers, maximum power point trackers, auxiliary battery chargers and DC to DC converters. A special section covers DC to AC inverters.

**Battery charge regulators and load controllers (off-grid systems only)**

Battery charge regulators (see Figure 18.1) are electronic devices which control the power output of the array so that it may not overcharge the battery. Sophisticated regulators trickle charge the battery when it is fully charged. They may also automatically compensate for temperature influences. Load controllers protect, on the opposite end, the battery from being completely discharged, by warning the user that the system is reaching a critical level or by cutting off the load. Sophisticated controllers are able to shed the least important loads when the battery cap-acity level is low. Many battery charge regulators have load controllers incorporated in them. These two electronic devices are essential in extending the lifetime of the battery component and special care must be given to choosing the right one for the application and **adjust it properly** before operating the system.

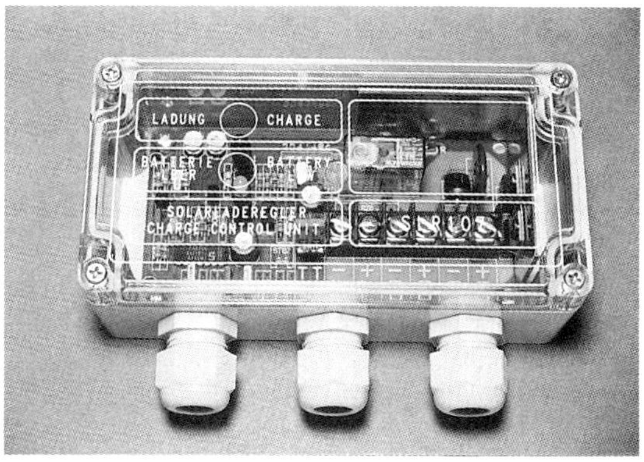


*Figure 18.1 Battery charge regulator. (Uhlmann Solarelectronic)*

The following guidelines can be useful to the designer in making a selection:

- Small PV systems with large battery capacity (15 times the maximum array current) may not need a battery regulator, the internal resistance of the battery will be sufficient to absorb the overcharge current. Note that, when overcharging liquid lead-acid batteries, gassing will occur. Special care must then be given to properly ventilating the battery and compensating for water loss periodically.

- Small PV systems with a less than 100 $W_p$ array may use a simple shunt regulator which will shunt the current from the array across the battery through a transistor. The advantage of this system is that there is no power loss when the battery is charging. This approach is not recommended for larger systems because the power to be dissipated through the transistor when the battery is full requires the use of a dummy load and can be potentially hazardous.

- For systems above 100 $W_p$, a series regulator should be used. This device opens the circuit between the array and the battery when the battery is fully charged. Large systems exceeding 30 A should have relays which can avoid forming electric arcs.

- Large PV systems should have regulators which can trickle charge the battery. It may also be necessary to divide the array in sub-arrays and use a charger with a number of relays to charge the battery. As the battery is charged, the regulator switches out the sub-arrays in a planned sequence.

- When batteries are exposed to a wide temperature range, the regulator/controller devices should automatically adjust the charging and discharging set points to compensate for the variable electric characteristics of the battery and protect it from damage.

**Maximum power point trackers**

Maximum power point trackers (MPPT) are electronic devices which optimize the output of the array or of a number of different sub-arrays to match the electric characteristics of the load. In a uniform array, MPPT are used with loads with variable optimum voltage such as water pumps. In an array with different or variable electrical configuration of sub-arrays, MPPT are used to match the power output of each sub-array to the load so that the energy output of the higher voltage sub-array does not reverse into the sub-array with the lowest voltage. MPPT must be chosen to fit the operating range of the load and its control points must be set carefully to ensure proper operation. Note that most grid-connected inverters have an integrated MPPT.

**Auxiliary battery charger (off-grid systems only)**

Auxiliary battery chargers are used in hybrid PV/generator systems to charge batteries by the generator when insolation levels are insufficient. The battery charger must be selected according to the power output of the generator and the battery charging requirements. The battery charger output must not exceed the maximum charging rate of the battery but it must be as large as possible in order to minimize the running time of the generator.

**DC to DC Converter (off-grid systems only)**

DC-DC converters are used where it is necessary to convert from one DC voltage to another when multiple loads with different DC voltage levels are present. A DC-DC converter should be selected according to the power requirement of the load requiring it. Consider that it may be cheaper to use AC equipment and one main inverter if there is more than one DC voltage level or if the power requirement is greater than 1 kW.

## 18.5 Selecting the inverter

Depending on the nature of the PV system, the inverter converts direct current from the array or from a battery bank to alternating current. In grid-connected systems the inverter permits PV-produced electricity to be fed to the utility grid. In stand-alone (off-grid) systems the inverter permits the operation of common AC appliances from a DC source. Both types of inverters are very different in design and operation and should not be mistaken for one another.

**Inverters for stand-alone systems with batteries**

Inverters for stand-alone PV systems on buildings are usually connected to a battery bank and their voltage input is relatively constant. This voltage is usually low, from 12 to 48 V, but depending on the load size the current requirement can be high. Since the power requirement of the inverter is driven by the load demand, the inverter must be chosen so that it can meet the maximum load demand and still remain efficient at the level at which it will be used the most. The following points should be addressed when choosing an inverter:

- The input voltage of the inverter must be rated to handle the full range of the battery voltage. A sensing circuit is useful to prevent damage when operation of the inverter is below or beyond its optimal operation points.

- The inverter should be rated at least 20% more than the maximum power requirement of the load to ensure that it can deliver this power for an extended time.
- Inverters for stand-alone applications can have different wave-form output quality. A cheap square wave inverter can be used for small resistive heating loads, hand tools or incandescent lights. Modified sine wave inverters are appropriate for most loads where harmonics in the wave form will not adversely affect the operation of the load (be aware that harmonics will add to the heat generated by power loss in a motor). Sine wave inverters can operate any AC load within their rated power range but are usually more costly.
- Inverters used with motorized appliances should be able to exceed several times their rated capacity for a few seconds to withstand the power surge during start-up. An automatic overload feature which disconnects the inverter after a few seconds is recommended to protect the equipment from failure.
- It may be useful to choose more than one inverter for an application with variable loads in order to have the inverter operating at its maximum efficiency for a range of power requirements. Multiple units connected in parallel (cascaded) to service the same load must be compatible and their frequency must be inter-regulated to be in phase. Some inverters can also produce three-phase AC output.

**Inverters for grid-connected systems**

Grid-connected inverters directly convert DC electricity from the PV array to AC electricity which is fed into the grid. These inverters must comply with strict grid output requirements so as not to destabilize the line or introduce parasitic harmonics. The wave form of the inverter must be an almost perfect sine-wave shape and its total harmonic distortion must be lower than 3-5% following the utility's specification. Their power requirement is dictated by the power output of the array and the inverter should be rated not more than about the maximum power output of the array in order to be in its most efficient range. Special care must be given so that a group of inverters operating on the same line are prevented from feeding electricity to the line if it fails (islanding phenomenon). Please refer to chapter 9 for more information on the different inverters for grid-connected systems.

The following recommended specifications can be used to select a grid-connected inverter:

- high conversion efficiency (> 92%);
- low start-up and shut-down thresholds;
- power factor > 0.85 (satisfies local utility requirements);
- low total harmonic distortion of output current (< 3%);
- maximum power point operation;
- current limiting function;
- low power consumption at night ($P_0 < 0.5\%$ of $P_{rated}$);
- automatic disconnect at utility fault conditions (deviation of V,f);
- automatic restart after fault is cleared.

## 18.6 Selection of the back-up system (off-grid systems only)

Selecting the proper diesel or gas driven generator depends on a number of electrical and physical factors.

Electrical factors include the demand characteristics of the load (load variation, peak demand, limits of operation, starting loads) and reliability of the system. Isolated systems will require a back-up system that can come online automatically and require little maintenance.

Physical factors include the location of the system (e.g. exposure to harsh environment), space limitation, noise protection, isolation requirements, engine cooling and ventilation, fuel availability and its storage, starting aids and maintenance requirements. Repair information on all major components of the engine should be readily available.

## 18.7 Selection of wire size and type

Selection of the proper wiring depends much on the design layout of the system and the different components being used. The following factors should be noted:

- In general, the DC portion of the system will be low voltage. Thus, DC wire runs should be kept as short as possible to minimize cost and voltage drop. Wire selection must be made on the basis of allowed voltage drop. The voltage drop in a cable must be small in order to ensure that power at the correct voltage is delivered to the load. Using electrical code safety standards to select the size of a cable is not sufficient to ensure that the voltage drop will be less than 2-5%.

  The following formula can be used to determine the voltage drop $\Delta V$ in a cable:

  $$\Delta V = R * 2L * I$$

  where

  | | |
  |---|---|
  | R | is the wire resistance in Ω/m, |
  | L | is the wire run (one-way) and |
  | I | is the current. |

  Please refer to Appendix III for selection of the proper wire size.

- In order to achieve the most economical wiring diameter, one can optimize the counter-current effects of cabling investment costs and PV power losses. The total costs are

$$K = k_L * A + \frac{I_{nom}^2}{g * A} * k_{PV}$$

In order to minimize costs, $\frac{dK}{dA} = 0$ or

$$0 = k_L + (-1) * \frac{I_{nom}^2}{g * A^2} * k_{PV}$$

Changing $\frac{I}{A}$ to current density we get the optimized current density as:

$$S_{opt} = \frac{\sqrt{k_l * g}}{k_{PV}}$$

Explanation of the symbols used in the formulae:

| | | |
|---|---|---|
| K | Total investment | [$/m] |
| $k_L$ | Cost of cabling | [$/mm$^2$/m] |
| $k_{PV}$ | Cost of the PV array | [$/W] |
| $I_{nom}$ | Nominal current | [A] |
| g | Conductivity of the cable | [m/Ωmm$^2$] |
| S | Current density | [A/mm$^2$]. |

An example calculation for the total costs is illustrated in Figure 18.2 which also clearly shows the optimum.

| Resistance of Copper Wire | | | | | | |
|---|---|---|---|---|---|---|
| **Cross Section (mm²)** | 1.5 | 2.5 | 4 | 6 | 10 | 16 |
| **Resistance R (Ω/m)** | 0.0137 | 0.0082 | 0.0051 | 0.0034 | 0.0019 | 0.0013 |

*Table 18.1 Resistance of copper wire.*

- Appliances should be arranged in separate circuits for effective management and identification. AC and DC circuits must be well identified.
- Flexible cables should be selected in all places which need a lot of work. Cables made with several strands are preferred because a single strand can break more easily after bending.
- All wiring must be done to outlast the life-time of the system. Proper care must be given to cable connectors which will not degrade or loosen with time.
- All wiring should be selected according to the environment it faces and the protection it receives from it. There are many types of wires available. Some may be buried directly in the ground, others may need to be protected against humidity, UV or overheating.

## 18.8 Specification of safety devices and switchgears

Circuit layout must ensure proper safety for the user, the maintenance crew and the equipment in all conditions of operation. In larger systems, circuit layout should include monitoring points to measure the perform-ance of the system at all times.

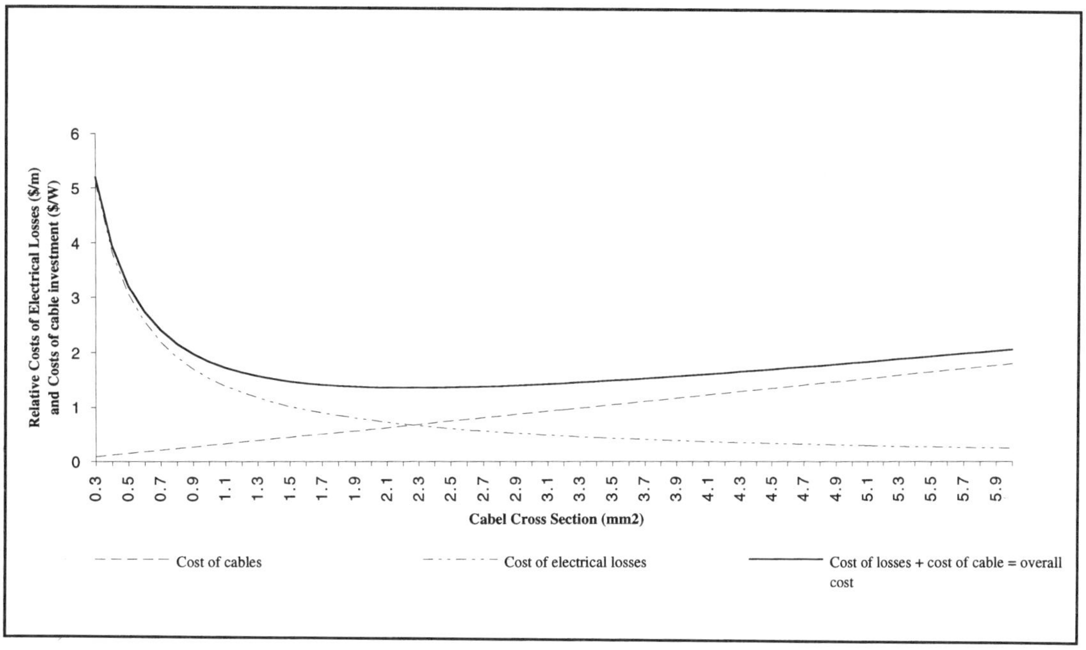


*Figure 18.2 Result of a total cost calculation for a typical string cable with a nominal current of 3.5 A ($k_L$ = 0.3 US$/mm²m, $k_{PV}$ = 7 US$/W).*

**Switches**

Each circuit must be protected against system-failure and short-circuits. The system itself must be rendered inoperational by cutting off a main circuit switch to allow work on the system without danger.

DC circuits must have properly rated DC switches. In switching DC, an arc is formed when a switch is opened and will last till the contacts are some distance apart. If the arc continues for long, it can burn the contacts and shorten their life. High current DC lines can create long arcs which will destroy the switch or the system and can be hazardous to the operator if they are not properly selected.

Switches with a built-in timing mechanism can be useful to cut off an appliance which may be left on inadvertently. Energy efficient techniques are a must for solar systems.

**Fuses and breakers**

Select fuses and breakers to protect the circuit from shorts and current surges. Each circuit should be properly laid out and connected to a main distribution box to minimize power loss and increase safety of the system.

**Surge protection**

Circuit layout must include protection of the circuit and of the user from surges derived from lightning and other transient inputs.

**Plugs and sockets**

Plugs and sockets should be different for DC and AC appliances. DC circuits should use plugs and sockets which identify the polarity of the conductors and prevent the appliance from being reverse connected.

## 18.9 Checklist of required parts

The following table lists the system parts and important features.

| Items | What to look for |
|---|---|
| **Array** | |
| Electrical Specs | $W_p$, $V_{OC}$, $I_{SC}$, I-V curve |
| Modules | Size, type modularity |
| Mounting frame | Material, facility of mounting, nuts and bolts |
| Mounting base | Cement, fastening system |
| Fence | Protection against vandalism |
| Connection | Facility of connection |
| **Batteries** | |
| Electrical Specs | Ah capacity, voltage, cycling, discharge capacity, charging threshold |
| Physical limitation | Size, weight, environmental protection needs, ventilation, frame |
| Electrolyte | Type, specific gravity, distilled water need |
| Connection | Ring clamps, grease |
| **Controller** | |
| Electrical Specs | Set points, adjustability, temperature compensation, upgrading |
| Protection | Fuse, reverse-polarity protection |
| Connection | Type, ease of connection |
| Options | Metering/monitoring capability, tamper-proof, alarm, computer-coupling |
| **Back-up generator** | |
| Electrical Specs | Type, VA, voltage, regulation, frequency, THD, surge capacity |
| Fuel | Diesel, propane, gasoline, availability, consumption |
| Physical limitation | Size, weight, engine cooling and ventilation, noise, heating requirements |
| Maintenance | Periodicity, ease, overhaul requirements |
| Options | Automatic starter, battery charger, shut-down protection |
| **Inverter** | |
| Electrical Specs | DC input voltage, AC output voltage, phase, frequency, THD, surge capacity, shape of curve, power consumption during standby, local code safety requirements |
| Physical Specs | Temperature limitation, humidity limitation, size, noise level, ventilation needs |
| Utility Requirements | Safety, power quality, trip limits, protection details |
| **Distribution** | |
| Cables | Distance, voltage drop, flexibility, environment |
| Conduit | Type, coupler, environment |
| Connectors | Crimp, ring, wiring terminals, splicing |
| Safety protection | Fuses, breakers, surge protection, main switch, circuit switches, grounding |
| Distribution box | Facility of wiring, upgrading, environmental protection, vandalism protection, grounding |
| Plugs and sockets | Polarity identification and protection |

*Table 18.2 System parts checklist.*

## Section E

# INSTALLATION AND MAINTANANCE

## Principal Contributors

**Chapter 19:** **Photovoltaic System Installation Guidelines** [1]

Peter Toggweiler (PMS, Switzerland)
Christian Roecker (EPFL-LESO, Switzerland)

**Chapter 20:** **Photovoltaic System Operation and Maintenance**

Alvaro Gonzales-Menendez (Ciemat-IER, Spain)
Jimmy Royer (SOLENER Inc., Canada)
Jyrki Leppänen (Neste, Finland)

**Chapter 21:** **Commissioning of Photovoltaic Systems**

Mats Andersson (Catella Generics AB, Sweden)

[1] Part of the text based on "The Design of Residential Photovoltaic Systems", volume 5 "Installation, Maintenance and Operation Volume", document number SAND 87-1951-5, edited by Dr. Gary Jones and Dr. Michael Thomas.
Contributors to this report were James Huning, John Otte, Elizabeth Rose, Russ Sugimura and Kent Volkmer of the Jet Propulsion Laboratory.

## Chapter 19

# Photovoltaic System Installation Guidelines

### 19.1 What the installer needs to know

The installation of a photovoltaic system in a residence should not be an unusually difficult task as long as the few unique characteristics of photovoltaics are well understood. This chapter describes the steps for a typical residential photovoltaic system installation and provides special information useful to the system installer.

Because the DC part of the photovoltaic system is 'activated' when modules are illuminated, there is the possibility of electric shock hazard with varying degrees of severity. Safety rules must be followed, and the entire work crew should be instructed about possible shock hazards and how to avoid them.

- PV modules generate electricity when exposed to light
- When wired, PV modules may generate a lethal shock
- A DC current spark is more dangerous than an AC spark

*Table 19.1 Instructions for work crew about PV unique hazards.*

### 19.2 Component delivery: inspect, test and protect

To minimize the possibility of theft, vandalism and other risks to handlers, a number of suggestions for storing and handling system components is provided. In addition, suggestions for inspecting and testing components prior to installation are given. Of course, these procedures may vary according to specific site situations and the local construction protocol. A large multiple-unit installation will require variations to the procedures suggested here for an individual residence.

#### 19.2.1 Timing is important - plan the delivery schedule

When a PV system is installed during a new home construction, system components should be ordered to arrive at the site at the appropriate time in the construction sequence. Coordination of building construction with array installation may be especially critical. For example, building codes usually require the roof installation to be completed before the electric wiring may be installed. When the array itself becomes the roof weather seal, as in the case of an 'integral' array design, array installation can obviously pace the remainder of the home construction. This consideration will be less important for array designs that do not constitute the roof weather seal.

- Cracked cover glass
- Bent or dented frame, if applicable
- Loose or broken wires
- Check all modules for proper open circuit voltage when exposed to full illumination

*Table 19.2 Inspection of PV modules upon receipt for obvious damage or defects.*

Delivery of all PV components at the same time will usually minimise the expense of return visits by installers and the expense of providing secure storage for valuable components awaiting installation.

### 19.2.2 Component function check prior to installation

Once the photovoltaic modules have been received, care must be exercised in removing them from their shipping containers. If the terminals are exposed, an electric shock is possible. Although the electric shock hazard from a single module is low, it may be sufficient to startle the person handling it and cause her or him to drop and damage the module.

Each module should be visually inspected for damage or obvious defects such as cracked glass or cells. Modules with cracked cover glass or cells, or with bent or dented frames, should not be installed in the array, but returned to the vendor for replacement. Similar visual inspection should be done for other system components as well, especially for the power conditioning unit. If the PCU is damaged during shipment, the manufacturer should be notified as soon as possible so that repair and/or replacement can be made. The damaged shipping container should be saved as evidence in the event that the unit must be returned.

A simple check of the module's electrical function is recommended as part of the receiving inspection. Because each module has been electrically tested by the manufacturer, sophisticated testing is not necessary at this stage. An I-V (current-voltage) curve is useful at the module level only if the design calls for matching array source circuit power outputs. A simple check of open circuit voltage ($V_{OC}$) will be sufficient. (Note: This check should be performed only with the module directly facing the sun.) The $V_{OC}$ check is a very rough test to detect only a serious fault or malfunction in the module. The measured value of $V_{OC}$, even for a properly functioning module, will vary according to its temperature and the intensity of the sunlight. A $V_{OC}$ reading of no more than 20 percent below the manufacturer's specified value will generally indicate that the module is functioning correctly. During extended cloudy periods, this check should be dispensed with since results will be too difficult to interpret. Additional electrical testing will be accomplished during the installation and checkout phases.

### 19.2.3 Protecting PV components

Photovoltaic system components represent a significant investment and value. Module and power conditioning units should be kept in a secure location to prevent theft or vandalism. In addition, the components should be protected from weather until installation. Modules should not be left in direct sunlight while resting in their open shipping containers. Photovoltaic modules stored in the sun can become extremely hot even if crated. If the installer removes hot modules from the packing crate without using protective gloves, a burn is possible and the module could be dropped and damaged. If possible, excessive stacking of modules, one on top of another, should be avoided. A preferred method is to store modules on their edges.

## 19.3 Array installation: the key step

Installation of a rooftop array of PV modules will be the most challenging part of residential system installation. A variety of mounting methods are available, which may be used with a number of roof structure styles and designs. Installation of the array requires mechanically mounting the modules, attaching the electrical interconnections, and checking the performance of completed array source circuits. All phases of array installation involve working with electrically active components. Consequently, all workers must be familiar with the potential hazards of installing PV and with the specific procedures appropriate to the system being installed. Each option for mounting and wiring an array will present its own special installation requirements. For this reason, the system designer must provide detailed installation procedures for each system design.

### 19.3.1 Array mounting options

A number of factors determine which specific array design is selected. Among these factors are:

- whether the system will be integrated into a new residence or an existing home (retrofit);
- type of roof construction;
- roof orientation;
- local climatic conditions;
- institutional factors, including local covenants and restrictions.

There are four basic mounting options: direct, integral, stand-off, and rack. In a direct mount, modules are mounted directly on the roof sheathing and shingles.

In an integral mount, the photovoltaic modules replace the roofing material, including sheathing. In this case, modules are mounted directly on the roof rafters and provide the weather seal. In a stand-off mount, the modules are mounted a few inches above and parallel to the roof surface. Finally, in a rack mounted array, a frame or rack is first installed to support the modules. The orientation of the rack can be selected for the most appropriate direction or angle to the sun.

In a new construction, the system designer can select the preferred mounting scheme, which may then influence the overall roof structural design. For example, if an integral mount is selected, rafter and spacing specifications must be made compatible with the proposed module size and attachment methods. With a retrofit, however, the existing roof structure will largely dictate which mounting schemes are feasible. A retrofit application will generally rule out an integral mount because of the cost or unsatisfactory rafter spacing. Or, in another case, an existing roof may depart sufficiently from optimal tilt or orientation such that only a rack mount may be used.

Each option presents special design constraints to the system designer. Because the integral mount becomes part of the roof itself, the builder must give particular attention to scheduling the PV installation, especially the wiring of the array. Such items as how and where to run the electrical wiring, where to place the roof insulation and where to situate the junction boxes, need to be clearly sequenced for the builder. A possible advantage to the other three mounting options - direct, stand-off, or rack - is that the builder has the option of scheduling array installation after the roof structure is completed. This approach may prevent interference between work crews when the array is not being installed by the general contracting crew.

To some degree, local building code requirements also help to determine which mounting options are feasible. The architect and system designer will have identified any local requirements before the final system design is selected. Factors normally addressed by codes include the extra weight and wind loads that modules and any support structure will add to the roof.

Regardless of the mounting option selection, the architect or system designer must give attention to the requirements of installers or service personnel in two particular areas:

- **Accessibility.** The mounting option must allow for safe and effective module removal or maintenance. Any special hardware requirements should be specified by the system designer. The mounting option must not only allow for physical accessibility, it must also allow access to the electrical termination on the back side of standard PV modules. The integral mount is usually superior, since this provides direct access to the back of the modules from inside the building. Direct or stand-off arrangements require partial removal of the module while working on the roof surface to gain electrical access. This operation may require special ladders or scaffolding to prevent standing and working directly on the array surface.

- **Weather sealing.** Any array mounting scheme must provide for, or at least not jeopardise, the roof weather seal. This is most important in the case of the integral mount. Proper attention must be given to casketing and hold-down materials, both during design and during actual installation.

No unusual construction practices are required, although with an integral mount special care must be taken to ensure a weather-tight fit. Experimental designs are satisfactory in this regard, although applying liquid sealants may require working in dry, above-freezing weather conditions.

### 19.3.2 Module attaching methods

The actual procedure used in securing the modules is determined, in part, by the specific array design, including the type of module and mounting option used.

Detailed installation procedures must be provided for each system design. The procedures should emphasise that DC shock is a potential hazard at all times, not only during the array wiring phase. Accordingly, protective equipment such as insulated gloves should be used whenever modules are handled or are being attached to the support structure. If feasible, the modules may be covered to block the light and preclude shock hazard.

Moderate to high velocity winds present difficulty in handling modules on a rooftop and may significantly affect worker safety and installation scheduling. For example, in areas where high wind velocities are a typical occurrence in the afternoon, module mounting may have to be scheduled for early morning or early evening.

A work crew consisting of at least two people should be used to install the modules. The modules should not be walked on at any time. Some experimenters have suggested three-person crews, two to handle the module and one to place it in the support structure. Lifting large modules or multimodule panels to the roof may require a hoist or crane.

With several experimental array schemes, mounting holes had to be drilled to allow the module to be attached to the support structure. Special care is required in measuring for the mounting holes. If one hole is misplaced and the module mounted, each subsequent module will be out of alignment. In some cases, the installer may prefer to drill mounting holes as each module is attached, so ensuring that no time will be wasted in having to redrill them.

On sloping roofs, precautions used by experienced roofers, such as a lifeline and safety belt may be required to prevent accidental falls. This is especially true for roofs that are steep or inherently slippery. Hard hats should be worn by those working on the ground during array installation. Non-conducting tools and other support equipment should be used.

### 19.3.3 Array wiring instructions

Because of the diversity of acceptable array electrical wiring systems and termination methods, it is not possible to describe an installation procedure that addresses every contingency. Therefore, it is especially important that the PV system designer provides detailed procedures and design-specific recommendations to facilitate proper and safe array wiring. This section discusses points that should be addressed in detailed procedures.

Because the wiring system and termination must be installed at the site, local building code requirements must be observed.

The various acceptable array wiring and termination methods require no special mechanical skills or tools beyond those normally found at a construction site. Generally, differences between PV and other wiring installation methods consist of the additional precautions required by PV due to its constantly energized operating characteristics.

Existing US National Electrical Code requirements specify that non-current-carrying conductive structures are to be solidly grounded. This is consistent with present design practice and, in general, should be one of the first steps of any PV electrical installation that utilizes a conductive mounting structure for the PV modules. The array design should also ensure that modules or panels with a conductive frame are effectively bonded to ground.

The remaining step in the electrical installation of a PV array consists of manually wiring a number of series-connected modules that will achieve system voltage. This series string of modules, which constitutes the source circuit, is then connected in parallel with other source circuits to form the system output.

Contact with conductors at a voltage greater than about 50 V can result in a lethal shock. In most cases, a residential PV system voltage will be substantially higher - in the order of 200 V. Since PV modules develop full voltage potential at low illumination levels, contact with conductors that connect a series of modules in a source circuit carries the potential of a lethal shock. Remember, unlike more conventional electrical sources, PV modules cannot be 'turned off'.

EXTREME CAUTION SHOULD BE EXERCISED WHEN WORKING WITH CONDUCTORS AT THE SOURCE CIRCUIT LEVEL

Two additional cautions must be emphasised. Because DC arcing is possible when a DC circuit is interrupted, caution must be exercised when a DC connection or termination is broken. Installation procedures should ensure that source circuits will remain open prior to circuit checkout so that no current will flow. If the source circuit is accidentally closed (e.g. by shorting across the terminals of the circuit disconnect switch, or between circuit wiring and the grounded array frame), any subsequent opening of the circuit will produce a large DC arc. (These conditions carry other safety hazards as well and must be avoided.) Even though the arc itself may not be hazardous, the worker could be startled sufficiently by it to fall and be injured.

The second caution is based on the limited availability of short-circuit current and on the problem that protective devices may not function in the presence of a ground fault. This unique characteristic coupled with the fact that the array is electrically energized by the sun demands extra care in installation to prevent shock or fire.

As installation proceeds, both array wiring and terminations must be tested and checked. Specific procedures for installation and checkout should be written for each wiring and termination method. In general, test and checkout of each source circuit should be performed as early as convenient during the installation sequence. Prior to source circuit checkout, proper installation of the structure grounding system should be verified.

During the actual electrical interconnection of modules, particular attention should be paid to maintaining the correct polarity. Any series safety devices (i.e. diodes, fuses) should be checked for proper installation (secure connection, and proper polarity where applicable).

As a minimum, a voltage check should be performed on each source circuit, as well as on the photovoltaic output circuit (the full array). All disconnecting means should be exercised to verify correct operation. This will normally involve only a simple voltage check on the downstream side of each disconnection switch in both the open and closed positions. If voltage remains on the downstream side of a source circuit or array disconnection switch after it has been opened, either the circuit has been wired incorrectly or the switch is faulty. If the circuit proves to be correctly wired, the disconnection switch should be removed and tested with an ohm-meter and replaced if necessary.

Because each specific design should include a set of detailed wiring procedures, only generic procedures have been discussed. The critical issue is that the designer must anticipate the wiring requirements and give the installer sufficient details so that the safety of installers, occupants, and the photovoltaic system is ensured.

### 19.3.4 Location of the power conditioning unit

The preferred location of the power conditioning unit will be dictated, in part, by physical characteristics of the PCU, including size, noise production and environmental operating requirements. The unit must be properly mounted to minimise audible noise and vibration, and it must be placed in a secure location to prevent tampering (see Table 19.3).

The location of the PCU in the building will also depend in part on specific site climate. In areas that experience a large annual or daily temperature variation, the PCU will have to be protected from the weather so that the specified operating temperature range is not exceeded. In areas that have an equable climate, the PCU may be located outside the building, but it must be protected by a weather-proof box. If located outdoors, the ventilation ports on the box may have to be screened to prevent small animals from taking up residence inside the box.

Each PCU has a specified operating temperature and relative humidity range, which should be indicated in the manufacturer's specifications and on the unit itself. The temperature range (min-max) must be determined and the unit located such that exposure to temperatures beyond its safe operating range (e.g. not below freezing or above 38°C) is avoided. All PCUs produce waste heat (as much as 5-10% of nominal power), which means that adequate ventilation must be ensured.

If temperature requirements dictate an inside-the-building location and the PCU under consideration produces noise, it should be located where noise will not be an environmental nuisance to the occupants. Again, the amount of heat produced must be taken into account if the unit is located in a small area (e.g. closet).

If the power conditioning design permits, attachment to wall studs is generally recommended. In this location it is more accessible for occasional monitoring, less accessible to small children, and less likely to hinder activities in the area. Manufacturers' recommendations should be observed regarding specific mounting methods. Some PCUs are designed for free standing and require no mounting procedures.

Lethal voltages are connected to the power conditioning unit, and it is imperative that individuals be protected from accidental contact with electrically active parts. For this reason, the unit is normally packaged in a cabinet that can be locked or otherwise secured. Other components of the power conditioning subsystem (e.g. an isolation transformer, if separate from the PCU) should be located within a lockable PV service box. 'DANGER HIGH VOLTAGE' labels should be placed on and near the power conditioning subsystem, regardless of its location.

## 19.4 Special protection devices

In addition to the array of PV modules and the power conditioner, several additional devices will be included in the system design to provide protection in case equipment should malfunction. Applicable codes and utility requirements dictate the protection devices that are required in the PV system.

- Choose location of PCU that meets temperature and humidity requirements
- Consider PCU-generated noise when selecting PCU location
- Ensure adequate ventilation since PCU produces significant heat
- Avoid public access to live parts

*Table 19.3 Location of Power Conditioning Unit (PCU).*

### 19.4.1 The 1984 US National Electrical Code

Specific recommendations of the US-NEC Article 690 are of general validity and include

- Use of overcurrent protection devices in all PV electrical sources, including the PV source circuit (string), PV output circuit (array), PCU output circuit, and storage battery circuit conductors.
- Protection against possible feedback of current from any source of supply, including the PCU.
- Inclusion of readily accessible, manually operable switch(es) or circuit breaker(s) to disconnect conductors.
- Surge protection and grounding methods.

### 19.4.2 Additional steps

Several additional steps are recommended to provide protection in the event of any malfunction. These include ground-fault protectors, alarms, system interlocks and disconnects, lockable boxes and special labelling, and specific placement of equipment. These items are not universally addressed in existing PV design guides.

### 19.4.3 Co-location of PV system components

To facilitate maintenance and protect service personnel, selected components of the PV system (e.g. blocking diodes, fuses, source circuit disconnects) should be co-located in a lockable DC service box. Co-locating equipment minimises wiring and the number of places service personnel must work. It also can provide a lockable barrier between high voltage equipment and untrained persons or children.

To minimise the shock hazard to service personnel during maintenance, it must be possible to electrically isolate the PV system from the utility. In addition, service personnel must be readily able to disconnect the array from the PCU. Lockable disconnects are important when maintenance must be performed on the roof array, away from disconnect switches. The power conditioner, fuses, switches, test points, alarms, and service interlocks may be conveniently located in a dedicated space.

A smoke detector may be used as an indicator of power conditioner faults or electrical wiring shorts that present a possible fire hazard. An ion-sensitive smoke detector is particularly recommended because of its extreme sensitivity. The recommended location for the smoke detector is directly above the power conditioner. The alarm must be audible within the occupied building.

Safety codes recommend an overcurrent protection device such as a fuse in every circuit. Overcurrent and overvoltage protection are now standard features built into today's power conditioners. However, a PV system represents a significant investment, and additional overvoltage surge arrestors may be cost-effective in certain cases.

It is generally agreed that a system cannot be economically protected from a direct lightning strike, although it may be protected against power surges from near strikes by the use of surge arrestors. Surge arrestors provide economical and effective protection and should be installed both line-to-line and line-to-ground. The surge arrestor should be installed on both the positive and negative legs of the array to protect the PCU from damage. Many people believe air masts and lightning rods provide little additional protection. Lightning surge protection may be unnecessary in areas where electrical storms are rare. Considering the value of PV system components and the low-cost protection provided by surge arrestors, their use is strongly recommended in all residential installations. The use of surge arrestors on the output of the PCU should also be considered to protect particularly sensitive house loads, such as computers, from occasional transients appearing in the PCU output.

### 19.4.4 Location of protection devices

The PV system or building design will frequently dictate the best locations for protection devices. Criteria to consider are:

- **Accessibility.** Protection devices must be accessible to service personnel so that they can be actuated (e.g. disconnect) or inspected (e.g. fuses) with a minimum of difficulty.

- **Wiring requirements.** A trade-off may be necessary because extremely accessible locations will generally require considerable additional wiring, and thus cause additional cost and failure risks.

Array and utility disconnects must be located as specified by local code, utility, or fire department requirements, usually in visible outside locations. The array disconnect is most commonly located near the power conditioner. A lockable disconnect should be used to prevent accidental closure during service. The array disconnect should be clearly labelled. In addition, fire department personnel should be notified that the disconnect only interrupts power supplied to the building, and that the roof array remains electrically active when illuminated. The AC disconnect should be located near the utility service entrance in a place accessible to utility personnel.

| |
|---|
| • Lightning surge protectors should be installed to protect the PCU on both AC and DC sides<br>• Locate grounding connection point for lightning power surge protection as close as possible to the device<br>• For proper foundation of the devices, they should be placed where the wire enters the building |

*Table 19.4 Location of surge protection.*

### 19.4.5 Labels, warnings and instructions

Warning labels and special instructions are important for the occupants' safety in case of a serious system malfunction. In addition, special labelling and instructions should be provided for service personnel and fire-fighters. A sign indicating the presence of a PV system should be placed on the utility pole or vault serving the PV building to alert utility service personnel.

Blocking diodes are used to prevent one source circuit from feeding current back into another source circuit in cases where one circuit voltage is higher than another. The system designer should consider locating these diodes in the PV service box where they will be accessible to service personnel. If diodes and diagnostic test points for each source circuit are located in the PV service box, necessary service can be more easily accomplished. Manual switches that allow each circuit to be independently disconnected (for testing) can also be co-located in the PV service box. Each switch should be clearly labelled to indicate which source circuit it disconnects. All circuits should be identified on the schematic diagram prepared for trained service personnel.

## 19.5 Inspection of the system

In most countries the newly installed PV system requires building code inspection by local officials. In addition to electrical inspection, structural inspection may be required. A residential photovoltaic system represents a new and unfamiliar technology to most inspectors in local agencies, at least for the near-term future. To preclude possible installation, checkout, and operation delays, necessary inspection requirements must be well understood and action must be taken to meet them at the appropriate times.

### 19.5.1 Inspection requirements

The electrical and structural inspection requirements vary according to the region of the country, both as a function of administrative jurisdiction (i.e. city or county) and specific utility.

The number of inspections and the level of detail to which a photovoltaic system will be inspected vary according to the region. In some areas, code requirements may be rigorous and followed closely, while in other areas requirements may be more lax. It is imperative that local requirements be clearly known before the design is completed and any inspection requirements be scheduled into the installation and checkout phases.

Because most inspectors will not be familiar with photovoltaic systems, especially the 'always-on' characteristic of the array, a thorough briefing of the inspector before a site visit may prove valuable. Informing the inspector about what will be seen, why PV differs from other electrical systems, and the protective features that are included in the system should reduce or eliminate any reservations or concerns the inspector may have due to the novelty of photovoltaics.

The local fire department station should also be informed about the existence of the photovoltaic system and the 'always-on' characteristic of the array. The fire department may not levy specific inspection requirements, but fire department personnel will probably appreciate knowing that a photovoltaic system is planned within their service area.

When residential photovoltaic systems are more common, many of the concerns noted above will disappear. Until then, the architect/system designer should take the initiative to alert regulatory and service agencies to the existence and unique properties of photovoltaic systems.

#### 19.5.2 Utility requirements

Utility requirements vary, but the system designer and builder can be certain that the utility wants assurances that the system will not feed poor quality power into the grid (i.e. power having excessive noise or harmonic content), and that it will shut down in the case of an emergency or loss of utility power. In the initial design stage, the system designer must check with the specific utility to determine its requirements. For the near-term future the utility may even ask to see PCU performance data, but as photovoltaics become more and more common, the utility will likely specify only generic requirements that the system designer must meet. The utility may require a complete single-line diagram delineating control, protection and metering functions as part of the application's procedure. In any case, consult utility representatives before finalising the system design.

### 19.6 Installation summary

Detailed instructions are required for installation personnel, particularly for mounting and electrically wiring the array. These instructions should cover the points outlined below:

#### 19.6.1 Preparation of system installation

- Work crew briefed on PV procedures and safety hazards
- Components delivery scheduled
- Components checked upon receipt:
  - Modules
    - No broken glass or cell
    - Frames straight and undented
    - Open circuit voltage ($V_{oc}$) approximately equal to design value
  - Power conditioning unit
    - No visible damage
- Components safely stored in weatherproof location until installation.

#### 19.6.2 Installing the array

- Integral mount:
  - Modules installed before house wired
  - Modules may be wired one at a time as installed, or after all modules are installed
  - Sealants may require dry, above-freezing weather
- Direct or stand-off mount:
  - Installed after roof is intact
  - Modules wired one at a time as installed

- Rack mount:
  - Installed after roof is intact
  - Modules may be wired one at a time as installed, or after all modules are installed
- Safety procedures include:
  - Non-conducting ladders or scaffolding
  - No walking on module surfaces
  - Two-person work crew as minimum
  - No module handling in high wind velocities
  - Safety line on steep roofs
  - Briefing on unique design aspects or electrical safety hazards
  - Prevention against falling down
- Electrical wiring and checkout:
  - Array frame grounded before module installation
  - Wiring method and schedule per designer's procedures
  - Open circuit voltage ($V_{OC}$) check on each source circuit as wiring is completed
  - No attachment of array to power conditioner prior to initiating system-level checkout
  - Array left in open circuit condition.

### 19.6.3 Installing the power conditioning unit

- Locate to satisfy temperature and humidity specifications, usually done by the designer
- When possible, stud mount
- Protect from tampering
  - Verify seal on locked enclosure or
  - Place in lockable service box or
  - Dedicated space
- Good access for maintenance
- Allow free air circulation
- Cable inlet from the bottom.

### 19.6.4 Installing system protection devices

- Observe requirements of
  - Local building or electrical codes
  - Local utility and fire department
- Recommendations include
  - Overcurrent protection devices in source circuit, array, PCU output
  - Feedback protection from any supply source
  - Accessible, manual disconnects on ungrounded conductors
  - Surge protection and ground
  - Smoke detector above PCU
- Co-locate circuit overcurrent protection devices with test points
- Labels, warnings, and posted instructions
  - AC and DC disconnects
  - System circuit diagram
  - Test point labels and appropriate readings
  - Warnings at points of electrical hazards
  - Sign on utility pole or vault.

### 19.6.5 Inspecting the entire PV system

- Brief inspectors and schedule in advance:
  - Building and electrical
  - Utility
  - Fire department.

It may be an advantage to install the inverter first and the connection to the grid as well as the DC-cabling. Finally the PV-array can be mounted and start-up procedure can follow immediately.

## 19.7 Installation examples

### 19.7.1 Structural glazing example

The photovoltaic modules are fully integrated in a commercial building in a rear-ventilated construction. Frameless modules fitted into the balustrades and voids of the window areas were used.

The prefabricated facade elements were built into the building within a single day.

Figures 19.1 through 19.4 show steps between beginning and end of installation (Solar-zentrum Freiburg, SST GmbH, Germany).

*Figure 19.1 Structural glazing: Building without facade.*

*Figure 19.2 Structural glazing: The first facade element is being mounted.*

*Figure 19.3 Structural glazing: Facade mounting continues*

*Figure 19.4 Structural glazing: The completed building with PV facade.*

### 19.7.2 Flat roof installation

This system is an example of how standard photovoltaic elements can be integrated into a flat roof covered by conventional concrete paving stones (50 cm x 50 cm).

A concrete base which occupies a surface equivalent to that of two paving stones (i.e. 100 cm x 50 cm) was developed especially for this purpose. Current ballast norms for PV installations were easily met by the weight of the concrete (60 kg). As this is a pilot installation, the PV modules were attached to the concrete blocks in two different ways.

Figures 19.5 through 19.14 show the construction stages: The gravel of the original roof was removed and paving stones were displaced to make space for the prefabricated concrete module bases (Figure 19.5).

*Figure 19.5 Concrete foundation blocks and plates installed*

It takes two people to move one of these concrete blocks. Once they are in their place (Figure 19.6), the photovoltaic modules can be mounted directly or indirectly:

**Direct fixing** of the modules to the base is achieved with a double-sided auto-adhesive tape which is applied directly to the concrete. Good weather is necessary for this three-step procedure: first the concrete is cleaned and primed (Figure 19.6), then the auto-adhesive tape is attached to the concrete and finally the PV modules are put into position and secured by pressure (Figure 19.9).

*Figure 19.6 Cleaning and priming of concrete block.*

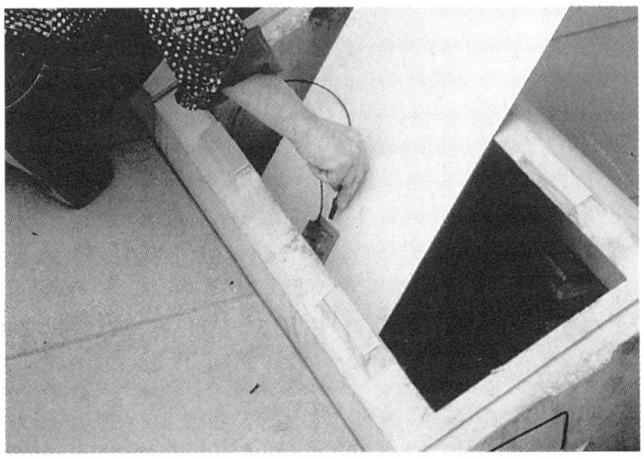

*Figure 19.8 Electrical wiring, using plugs, before gluing the module.*

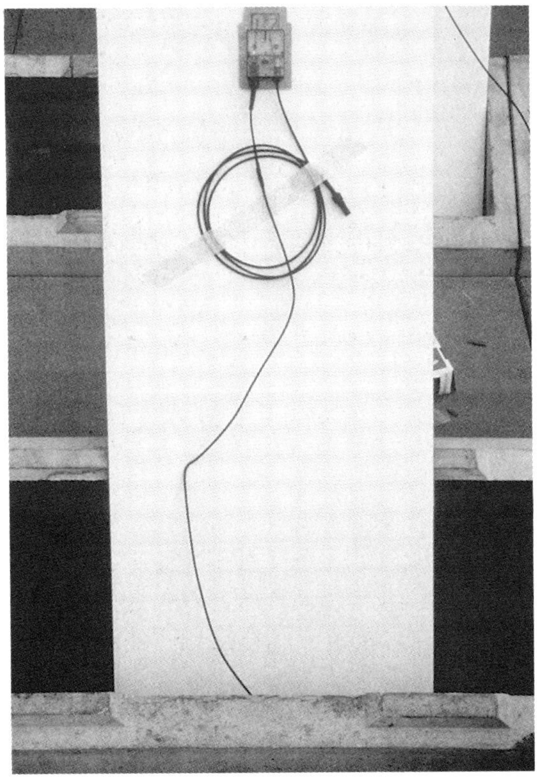

*Figure 19.7 PV module for direct mounting, ready with connection box.*

*Figure 19.9 Gluing of module to concrete block.*

**Indirect fixing** requires stainless steel brackets, which are glued to the module beforehand (Figure 19.11) and then attached to the concrete base as shown in Figure 19.12.

*Figure 19.10 Bracket position on the concrete block.*

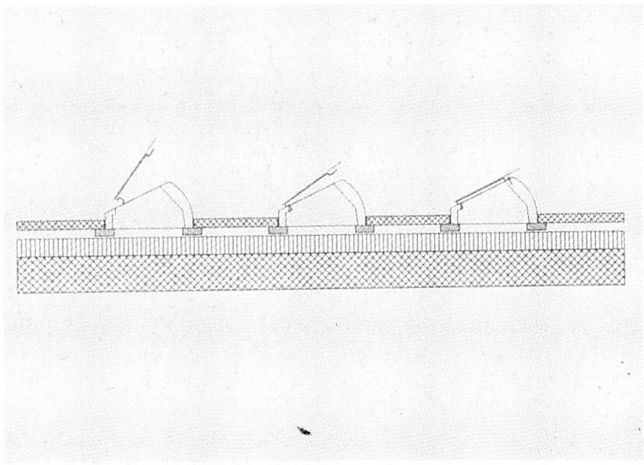

*Figure 19.12 Attaching PV modules with brackets to concrete blocks.*

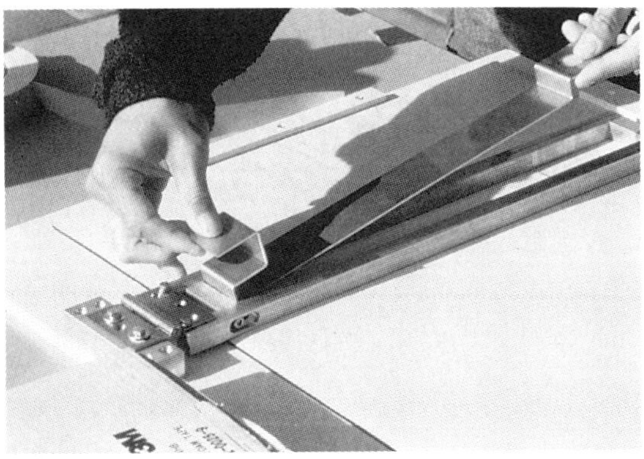

*Figure 19.11 Bracket gluing to PV module, with a template.*

*Figure 19.13 The completed flat roof installation.*

Both mounting techniques are accompanied by a very simple and efficient wiring system. It is arranged around a connection box with integrated diodes and connectors, which is glued to the back of the PV module (Figure 19.7). When the modules are mounted, the wires between them are laid underneath the concrete bases and simply connected by plugs (Figure 19.8). Thus, the wiring is protected from mechanical damage as well as from UV rays etc. All wires enter the building at the same place. Inside, they are connected to the inverter which feeds the electricity directly into the grid.

This system integrates very elegantly into the building. At the same time it allows easy access for roof maintenance and enough space for people to pass (Figure 19.13).

## Chapter 20

# Photovoltaic System Operation and Maintenance

### 20.1 Introduction

Properly designed and installed PV systems require little maintenance due to the absence of moving parts and to the high reliability of their components. Nevertheless periodic maintenance is recommended in order to ensure a good performance of the PV system and extend its useful life.

Although in general all PV systems have similar elements and/or subsystems (PV modules, battery, backup generator, regulator(s), inverter(s), cabling etc.), the level or type of maintenance depends on the complexity of the system and its use. In this chapter the maintenance of PV systems in buildings is discussed. A collection of procedures to be carried out periodically is suggested and described. Since it is good practice to establish a site log to record all maintenance schedules and comments, examples of log reports are presented.

*Figure 20.1 Back-side view of a PV array.*

### 20.2 PV array maintenance

The PV array (see Figures 20.1 - 20.3) consists of PV modules, cabling and a mounting structure. This subsystem is very reliable and does not require intensive maintenance.

Dust accumulated on a PV array can decrease its efficiency and performance. Normally rain will be enough to clean the modules if it is not a very dry site.

But in urban or industrial areas or near busy roads where ordinary dust is mixed with heavier and greasy particles, cleaning using detergents is recommended from time to time (see Figure 20.4). Do not use abrasive brushes and/or foams. In most cases a simple water jet from a garden hose mixed with detergent will be sufficient.

Twice per year (following seasonal climate changes) it is recommended to:

- review the structure of the PV array ;
- check PV modules for cracked cells and glazing, delamination and cell interconnect corrosion ;
- check power output of individual array strings;
- check wiring, connection boxes for cracking, rodent damage, fraying etc.;
- check electrical leakage to ground.

*Figure 20.2* *PV array roof-integrated.*

For cabling and connection boxes of the PV array a quarterly visual inspection is recommended. Replace every box that shows some sign of corrosion or degradation. In addition, the lightning protection (varistors) must be checked regularly especially after stormy weather. Normally varistors have an indicator, which changes colour or position when the protection has worked due to lightning causing a higher voltage than its nominal value. In such a case, replace it as soon as possible.

A log sheet for maintenance of the PV array is provided in Appendix V.

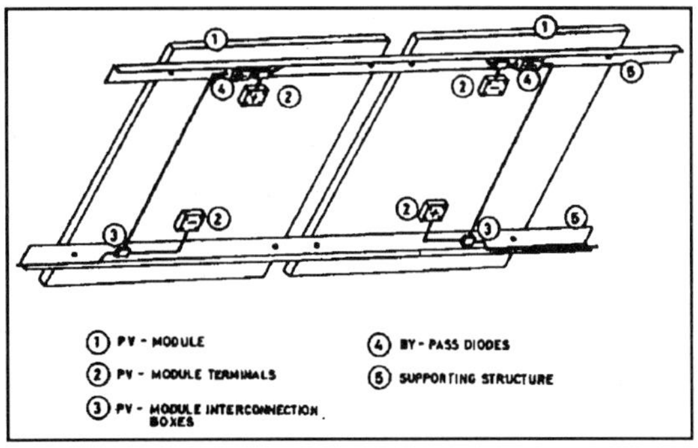


*Figure 20.3* *Detail of PV modules connection (back-side).*

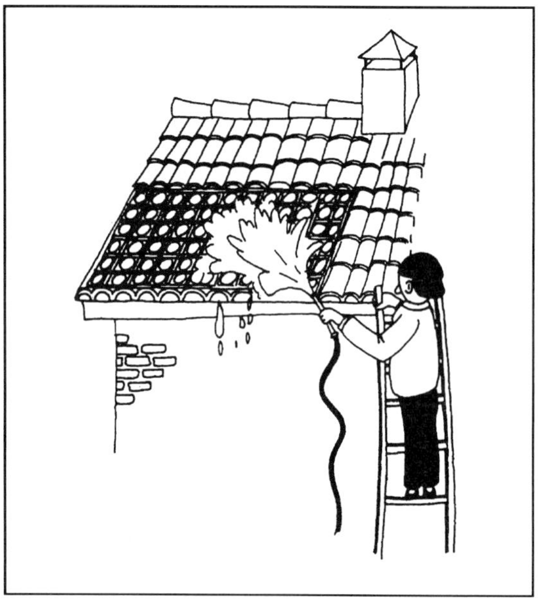

*Figure 20.4* *PV modules cleaning.*
*Attention: Short-circuit the output of the PV array through its corresponding switch before cleaning.*

## 20.3 Battery maintenance

### 20.3.1 General

The most critical subsystem in a PV-system is the battery. Depending on its maintenance, the useful life of a stationary battery for PV applications can vary from a few years up to ten or even more years. Therefore it is essential that preventive care of the batteries is done by the user and once a year by a specialist.

The following maintenance procedures are valid for lead-acid batteries, the most common in photovoltaic systems. For nickel-cadmium batteries the specific gravity measurements described cannot be used.

### 20.3.2 Routine battery tests

A routine test procedure needs to be set up before connecting the battery. The user should check the battery bank and look for the following points:

- differences in colour;
- sediment in cell boxes;
- corrosion on cells and connectors;
- cell cracks.

Normally, when the elements are connected for the first time, these points are checked out by the installer and/or manufacturer. If any of these signs is observed, the elements should not be connected. The user should recheck these points periodically and contact a specialist, if some of the mentioned signs are observed.

In case of corrosion of connectors it is recommended to clean them immediately. Corrosion of the terminals is indicated by a growing white powder (lead sulphate) on them. Each terminal must be cleaned very well with water and a small brush after the battery has been disconnected from the PV array and the loads. After reconnecting the terminals they are coated with grease or petroleum jelly. Finally the battery is reconnected through the corresponding switch to the PV array and the loads.

The state-of-charge of the battery can be checked on a routine basis by reading its voltage (see Figure 20.5). For a reliable reading, the battery voltage should be measured at the same time of day, usually just after sundown when the array is not charging and when the load is constant from day to day.

This voltage should be recorded in a log book. Action must be taken, if the state-of-charge of the battery is continuously at a low level. Continued use of a partially charged battery will shorten the overall life of the battery.

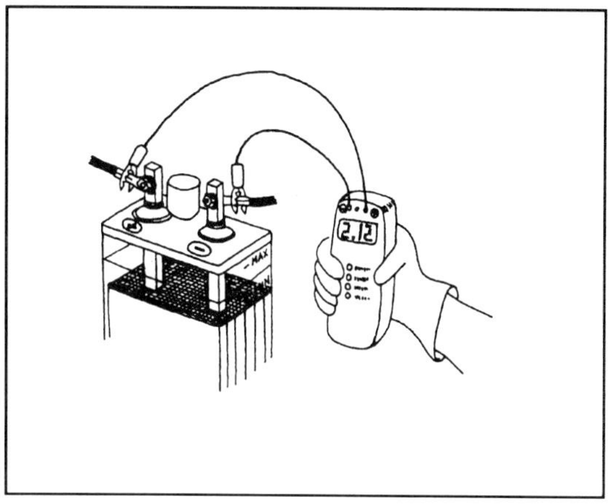


*Figure 20.5 Measurement of cell voltages. Note: For lead-acid elements and a temperature of 25°C, average voltages of 2.12, 2.07, 2.0 and 1.95 V/cell mean 100, 80, 60 and 40% of their rated capacities, respectively (details are given in Figure 7.4).*

### 20.3.3 Monthly battery tests

The user and/or a specialist must check the electrolyte level of the battery cells on a monthly basis and, if necessary, refill them using distilled water. Remember to avoid overfilling. Simultaneously it can be checked whether some cells have a lower level than other ones, which could mean an electrolyte leakage.

It is recommended to measure and record the cell voltages with a DC volt meter during the first year of service and to observe whether there are large voltage differences between the cells. This would indicate a defective and/or improperly connected cell.

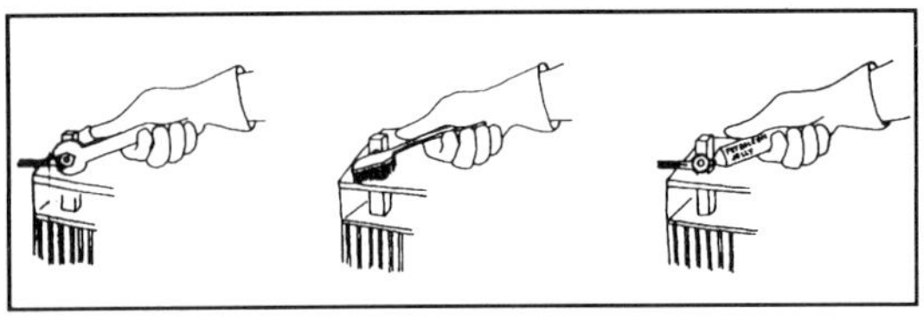

1) Disconnect the terminals
2) Clean them with a brush
3) Reconnect and fasten the terminals
4) Coat them with grease or petroleum jelly

*Figure 20.6 Battery terminals cleaning. Attention: Before cleaning, disconnect the battery from PV array and loads. The use of insulating gloves is always recommended.*

1) Always use protective gloves and proper labour wear
2) Always use distilled water
3) Refill slowly
4) Never overfill

*Figure 20.7 Battery refilling.*

### 20.3.4 Quarterly battery tests

Quarterly tests consist of precise measurement and record of cell voltages, specific gravities and temperatures, which are then used to estimate the working conditions of the battery.

For specific gravity measurements, a hydrometer is used. The hydrometer is inserted into each cell of the battery and measures the specific gravity of the electrolyte within the cells. At 15°C, values of 1.28, 1.25, 1.22 and 1.19 g/cm$^3$ would roughly mean states-of-charge of 100, 80, 60 and 40%, respectively. Note that temperature has an effect on the reading.

For batteries with liquid electrolytes it is also recommended to carry out an equalizing charging of the battery. With an equalizing charge each quarterly period, the stratification and sulphate problems will be corrected. The sulphate crystals are broken and the electrolyte circulates in the cells when the electrolyte bubbles during the last steps of charging. After each equalizing charge it is very important to check the electrolyte level and refill with distilled water if necessary (see Figure 20.7).

A typical equalizing charge voltage is 2.4 or 2.5 V/cell at 25°C during a minimum of 3 h and a maximum of 10 h charging. To do so, it is convenient to use an auxiliary energy source such as a backup generator (diesel or gas). It can also be done by connecting the PV array directly to the battery and thus bypassing the charge regulator, which normally cuts off the charging process before bubbling. Obviously, this can be done with the PV array only if there is enough solar radiation. This procedure requires special care and watching during the process.

## 20.4 Backup generator

Typical backup generators may be fuelled by one of a variety of petroleum based fuels such as gasoline, propane, natural gas, and diesel. Diesel generators (an example is shown in Figure 20.8) are the most common choice for applications where extended operating periods (greater than four hours per cycle) are required.

Some general maintenance procedures to be carried out after generator installation are listed below. **Note that these are general recommendations only. Consult the Operator's Manual or equipment supplier for a more complete listing of maintenance requirements.**

Maintenance procedures:

- Check engine oil level daily and/or before start-up.
- Change engine oil and oil filter after 25 hours of operation or at the end of the manufacturers recommended break-in period.
- Thereafter change engine oil and oil filter after every 50 to 150 hours of operation depending on engine type and operating environment (see Operator's Manual).
- Periodically check all fluid levels and add or replace per the manufacturers' recommendations.

A properly tuned generator will be able to operate electrical loads to its peak output rating while maintaining voltage and frequency output characteristics to within 5% of specification. Units unable to meet or exceed these specifications are typically in need of a tune-up and/or overhaul of the engine or generator component. This work should be carried out by qualified technicians or manufacturers' representatives.

*Figure 20.8 30 kVA diesel engine with automatic starting and stopping installed in a milking farm - hybrid PV/diesel powered.*

## 20.5 Power conditioning system maintenance

Maintenance of power conditioning equipment (regulators, inverters and controllers) should not be done by the user. Generally, its solid-state design makes it relatively maintenance-free but also difficult for anyone other than a trained person to repair. A simple visual examination from time to time by the user should show whether the equipment is performing well. From a user point of view it is recommended that the power conditioning equipment of a PV system should show: the instantaneous DC and AC voltages and currents, the total and partial (another counter with reset) amount of Ah generated and consumed and some visual and/or acoustic alarm in case of bad system performance, i.e. battery discharged more than 50% (in case of lead-acid batteries), too high or low battery voltage, frequency out of the required limits, etc. A check on proper operation of the control and alarm level must be done quarterly.

*Figure 20.9 Partial view of a power conditioning system installed.*

At present, most inverters include protection against overloads and they disconnect automatically. If the problem was just a temporary overload, a simple reset after disconnecting the offending load will be enough to solve it.

Above all, avoid any short-circuit of the AC output of the inverter during handling. When repairing or cleaning the inverter, disconnect the DC side after switching off the inverter. After a waiting period to allow for discharging of the capacitors (half a minute) disconnecting of the inverter may proceed.

## 20.6 Other subsystems

Apart from the PV array, the battery bank, the backup generator and the power conditioning equipment, there are other important items in a PV system, which need to be checked regularly, such as grounding, cabling, protection against atmospheric discharges etc.

For grounding a quarterly measurement of its resistance carried out by a specialist is recommended. Required values of grounding resistance have to be lower than 10 ohms. If the measurement obtained is higher than this value, it is recommended to irrigate the ground terminal. In case of very bad grounding values, it may be necessary to dissolve some salt in the irrigation water.

2

Almost all maintenance procedures described in this chapter can be carried out by the user. Although these procedures are easily done, it is very important to follow some safety measures to avoid accidents.

In order to avoid dangerous electrical shocks the user should cover the PV modules with an opaque blanket and disconnect the switch to the PV array from the battery before attempting to work on any circuit of the system. For larger arrays, the output of the array can be short-circuited.

When working with the battery, do not forget to wear safety boots and insulating gloves when you handle the battery bank, as well as appropriate protective clothing to prevent electrical shocks and burning by acid spills. Beware of shorting the battery terminals with tools and wrist watches. Short-circuits may not cause electric shock but they can cause arcing and burning. Do not smoke near the batteries because it can cause explosions due to the possible high concentration of hydrogen in the battery room. Battery room lighting should use anti-deflagrating lighting equipment.

Chapter 21

# Commissioning of Photovoltaic Systems

## 21.1 Commissioning of PV systems in buildings

After the installation of a PV system has been completed, it is important to go through some kind of acceptance procedure to assure that the installation work has been properly done and that the system is working correctly.

A complete commissioning procedure would have to cover a long list of items including:

- engineering review of system drawings;
- visual inspection of installation;
- mounting procedure (aesthetics, safety etc.);
- initial electrical performance tests on array, batteries, inverter, gensets etc.;
- safety for the user (insulation, grounding, earth fault protection etc.);
- safety for the system (overvoltage protection, fuses etc.);
- control of monitoring system (accuracy of sensors etc.);
- energy meters;
- overall system performance.

Most of the listed items are general subjects that apply to most electrical installations. Therefore, not every item is covered very deeply in this chapter; the electrical performance of the PV system, however, is different from other systems and it is discussed in more detail.

## 21.2 System configurations

A grid-connected PV system consists of the PV array, a mounting structure, inverter, junction box (one or several boxes containing diodes, fuses, lightning protection etc.) and cables. The utility interface for connecting the inverter to the utility network is also an important part of a grid-connected PV system. Fuses, an AC switch and one or two energy meters are the main parts of the utility interface.

A stand-alone system may have an inverter also but differs in the sense that it also consists of batteries and a charge regulator. In some cases, DC/DC converters are used to power appliances at other voltages than the system voltage.

In a hybrid system, which is a special type of stand-alone system, a genset and sometimes also a wind generator are parts of the system.

## 21.3 PV array

Even during the installation of a system it is recommended to check that all modules are connected to the junction box and that the open circuit voltage ($V_{OC}$) and the short circuit current ($I_{SC}$) of modules or strings are correct. It is always easier to find and repair a bad cable joint or take care of other failures before the whole system is completed. But even when the $V_{OC}$ and the $I_{SC}$ seem to be right one can find that the values around the operation point are out of the expected limits. A number of cells over even modules in a series string can be

# Recommended Reading

## Section A General

Canadian Photovoltaic Industries Association, *Photovoltaic Systems Design Manual*, CANMET-Energy, Mines & Resources Canada, Canada, 1991.

Commission of the European Communities, Directorate-General for Energy, *Community Energy Technology Projects in the Sector of Solar Photovoltaic Energy*, Athens, Greece, 1992.

Davidson, J., *The new solar electric home, the photovoltaics how-to handbook*, Michigan, Aatec, USA, 1987.

Derrick, A., Francis, C., Bokalders, V., *Solar Photovoltaic Products - a guide for development workers*, Intermediate Technology Publications Ltd., London, UK, 1991.

Duffie, J.A., Beckman, W.A., *Solar Engineering of Thermal Processes*, 2nd edition, John Wiley & Sons, New York, USA, 1991.

Fowler Solar Electric, Inc., *The Solar Electric Independent Home Book*, Worthington, MA, USA, 1991.

Green, M.A., *Solar cells, operating principles, technology and system applications*, Prentice-Hall, New York, USA, 1982.

Imamura, M.S., Helm, P., Palz, W., Stephens, H.S. & Associates, *Photovoltaic System Technology. A European Handbook*, Commission of the European Communities, Bedford, UK, 1992.

Johannsson, T.B. (Ed.), *Renewable Energy, Sources for Fuels and Electricity*, Island Press, Washington, DC, 1993.

Landolt-Börnstein, *Numerical Data and Functional Relationships* in *Science and Technology. New Series*, Vol. 4c, Climatology, Part 2, Berlin, Germany, 1989.

Lasnier, F., GanAng, T., Hilger, A. (Ed.), *Photovoltaic Engineering Handbook*, IOP Publishing, Philadelphia, PA, USA, 1990.

Maycock, P.D., Stiewalt, E.N., *A Guide to the Photovoltaic Revolution. Sunlight to Electricity in one Step*, Rhodale Press, Emmaus, PA, USA, 1985.

Palz, W. (Ed.). *European Solar Radiation Atlas*, Second Improved and Extended Edition. Volume 1, *Global Radiation on Horizontal Surfaces*. Volume 2, *Global and Diffuse Radiation on Vertical and Inclined Surfaces*, Verlag TÜV Rheinland GmbH, Köln, Germany, 1984.

Parker, B.F. (Ed.), *Solar Energy in Agriculture*, Elsevier, Amsterdam, The Netherlands, 1991.

Roberts, S., *Solar electricity, A practical guide to designing and installing small photovoltaic systems*, Prentice Hall, New Jersey, USA, 1991.

Roth, W., Schmidt, H. (Ed.), *Photovoltaik-Anlagen, Begleitbuch zum Seminar "Photovoltaik-Anlagen"*, Freiburg, Germany, 1994.

Sandia National Laboratories, Science Applications, Inc., *Design Handbook for Photovoltaic Power Systems*, McLean, USA, 1981.

Strong, S., *The Solar Electric House,* Chelsea Green, White River Junction, Vermont, USA, 1991.

Strong, S.J., Scheller, W.G., *The Solar Electric House. Energy for the Environmentally-Responsive, Energy-Independent Home*, Massachusetts, USA, 1991.

Warfield, G., *Solar electric systems*, Washington, USA, 1984.

Zweibel, K., Hersch, P., *Basic photovoltaic principles and methods*, Van Nostrand Reinhold Company, New York, USA, 1984.

## Section B Components

Brinner, A. et al., *Operation of a PV electrolysis system with different coupling modes*, Proceedings 9th EC PV Solar Energy Conference, Freiburg, Germany, 1989.

Buresch, M., *Photovoltaic Energy Systems*, McGraw-Hill Book Company, 1983.

Canadian Photovoltaic Industries Association, *Photovoltaic Systems Design Manual*, CANMET-Energy, Mines & Resources Canada, Canada, 1991.

CIEMAT PV-COURSE, *Energía Solar Fotovoltaica*, Madrid, Spain, June 1995.

Donepudi, V.S., Pell, W. Royer, J.M., *Storage Module Survey, IEA-Solar Heating and Cooling Program, Task 16-Photovoltaic in Buildings, Working Document*, Ottawa, Canada, October 1993.

Dunselman, C.P.M., van der Weiden T.C.J., van Zolingen, R.J.C., ter Heide, F.J., de Haan, S.W.H., *Design Specification for AC modules* (in Dutch), Ecofys report nr. E265, Ecofys/R&S/Mastervolt/ECN, Utrecht, The Netherlands, 1993.

Gabriel, C., TU-Graz, Wilk, H., *A simulation program for grid-connected PV plants*, Linz, 1993.

de Haan, S.W.H., Oldenkamp, H., Frumau, C.F.A., Bonin, A.; *Development of a 100 W resonant inverter for AC modules,* Proc. of 12th European Solar Energy Conference and Exhibition, Amsterdam, The Netherlands, April 1994

Haas, R., *The Value of Photovoltaic Electricity for Society*, Solar Energy, Vol. 54, No.1, pp. 25-31, 1995.

Häberlin, H., *Photovoltaik*, Aarau, Switzerland, 1991.

Hill, M., McCarthy, S., *PV Battery Handbook*, Hyperion Energy Systems Ltd., Ireland, April 1992.

Linden, Ed., *Handbook of batteries and fuel cells*, McGraw-Hill Inc., USA, 1984.

Panhuber-Fronius, C., Edelmoser, K., TU-Wien, *Resonant Concept for the Power-Section of a Grid-Coupled Inverter (2 kW)*, Proc. of 12th European Solar Energy Conference and Exhibition, Amsterdam, The Netherlands, April 1994.

Pletcher, D., Walsh, F., *Industrial electrochemistry*, Cambridge, UK, 1990.

Russell, M.C., *Residential photovoltaic system design handbook*, MIT, Massachusetts, USA, 1984.

Sandia National Laboratories, *Stand-Alone Photovoltaic Systems, A Handbook of Recommended Design Practices*, US Department of Energy, USA, March 1990.

Schaeffer, J., *Alternative Energy Sourcebook*, Real Goods Trading Corporation, California, USA, 1992.

Schmidt, H., *Der Charge Equalizer - die Lösung eines alten Batterieproblems*, Tagungsband Achtes Nationales Symposium Photovoltaische Energiesysteme, Staffelstein, Germany 1993, pages 229-239.

Strong, S., *The Solar Electric House*, Chelsea Green, White River Junction, Vermont, USA, 1991.

Wiles, J.C., *Photovoltaic Power Systems and the National Electric Code*, Southwest Region Experiment Station, New Mexico State University, Las Cruces, New Mexico, 1991.

Wilk H., *40 kW - Photovoltaic System with IGBT Inverter on the Soundbarriers of Motorway A1, Seewalchen, Austria*, 11th European Photovoltaic Solar Energy Conference, Montreux, Switzerland, October 1992

Wilk H., *200 kW Photovoltaic Rooftop Programme in Austria*, ISES World Congress, Budapest, Hungary, August 1993.

Wilk, H., *Solarstrom Handbuch 1993*, ARGE Erneuerbare Energie, Gleisdorf, Austria, ISBN 3-901425-01-2.

Wilk, H., *200 kW PV Rooftop Programme in Austria, First Results*, Proc. of 12th European Solar Energy Conference and Exhibition, Amsterdam, The Netherlands, April 1994.

Williams, A.F., *The handbook of photovoltaic applications, building applications and system design considerations*, Georgia, USA, 1986.

## Section C Architectural Integration

Davidson, J., *The new solar electric home, the photovoltaics how-to handbook*, Michigan, Aatec, USA, 1987.

Fowler Solar Electric, Inc., *The Solar Electric Independent Home Book*, Worthington, MA, USA, 1991.

Hullmann, H., Institut IB GmbH (ed.), *Ideenwettbewerb 1994: Photovoltaik in Gebäuden*, Hanover, Germany, 1994.

Humm, O., Toggweiler, P., *Photovoltaics & Architecture*, Basel, Switzerland, 1993.

Kiss, G. et al., *Building-integrated photovoltaics: a case study*, New York, 1994.

Kiss Cathcart Anders Architects, *Building-integrated photovoltaics*, New York, 1993.

Proceedings of the IEA international workshop, *Mounting Technologies for Building In-*

*tegrated PV Systems,* Mönchaltorf, Switzerland, 1992.

Schoen, A.J.N., (ed.), *Novem /IEA Architectural Ideas Competition "Photovoltaics in the Built Environment" - book of results,* Novem/Ecofys, The Netherlands, 1994.

Strong, S., *The Solar Electric House,* Chelsea Green, White River Junction, Vermont, USA, 1991.

Williams, A.F., *The handbook of photovoltaic applications, building applications and system design considerations*, Georgia, USA, 1986.

## Section D System Design

Canadian Photovoltaic Industries Association, *Photovoltaic Systems Design Manual,* CANMET-Energy, Mines & Resources Canada, Canada,1991.

Chapman, R.N., *Sizing Handbook for Standalone Photovoltaic/Storage Systems*, Sandia National Laboratories, Albuquerque, USA, 1987.

Davidson, J., *The new solar electric home, the photovoltaics how-to handbook*, Michigan, Aatec, USA, 1987.

Jantsch, M. et al., *Die Auswirkung von Neigungswinkel und Spannungsanpassung des PV-Generators auf die Leistung von PV-Systemen - eine experimentelle Untersuchung*, Tagungsband Siebtes Nationales Symposium Photovoltaische Solarenergie, Staffelstein, Germany, 1992, pages 217-226.

Kiskorski, A.S., *Power Tracking Methods in Photovoltaic Applications*, Proceedings PCIM '93, Nürnberg, Germany, 1993, pages 513-528.

Sandia National Laboratories, Science Applications, Inc., *Design Handbook for Photovoltaic Power Systems*, McLean, USA, 1981.

Sandia National Laboratories, Photovoltaic System Design Assistance Center, *The Design of Residential Photovoltaic Systems* (10 volumes), Albuquerque, USA, 1988.

Strong, S., *The Solar Electric House,* Chelsea Green, White River Junction, Vermont, USA, 1991.

Williams, A.F., *The handbook of photovoltaic applications, building applications and system design considerations*, Georgia, USA, 1986.

## Section E Installation and Maintenance

Canadian Photovoltaic Industries Association, *Photovoltaic Systems Design Manual*, CANMET-Energy, Mines & Resources Canada, Canada,1991.

*Maintenance and Operation of Stand-Alone Photovoltaic Systems*, US Department of Energy, Sandia National Laboratories, Albuquerque, USA, December 1991.

Strong, S., *The Solar Electric House,* Chelsea Green, White River Junction, Vermont, USA, 1991.

# APPENDICES

## Principal Contributors

**I Solar Insolation Data**

Kimmo Peippo (Helsinki University, Finland)

**II System Sizing Worksheets**

Jyrki Leppänen (Neste, Finland)

**III Wire Sizing Tables**

Jimmy Royer (SOLENER Inc., Canada)

**IV Tender Documents**

Peter Toggweiler (PMS, Switzerland)

**V Maintenance Logsheets**

Alvaro Gonzales-Menendez (Ciemat-IER, Spain)

**VI Trade-Off Considerations**

Oyvin Skarstein (Inst. of Technology, Norway)
Kimmo Peippo (Helsinki University, Finland)

**VII Cost of PV**

Heribert Schmidt (Fraunhofer ISE)

**VIII Glossary**

(several)

## Appendix I

# Solar Insolation Data

The following tables contain long-term monthly and annual averages of daily solar insolation (in kWh/m$^2$, day) on horizontal surfaces in selected locations, as well as descriptive temperature data, compiled from various sources. As an indication of the effect of the array inclination and orientation and possible shading, monthly and annual conversion coefficients are computed from horizontal to inclined surfaces for four latitudes: Sodankylä, Finland (67.5°N), Copenhagen, Denmark (55.7°N), Madison, USA (43.1°N) and Phoenix, USA (33.4°N), to be used for estimating the available insolation in accordance with the design procedure (i.e. a figure 0.5 appearing in the table indicates that the corresponding monthly or annual insolation is 50% of value for horizontal level). The coefficients have been calculated using a numerical simulation program PHOTO and the hourly meteorological Test Reference Years of the locations (except for Phoenix, where synthetic data were used). The numerical conversion tables are visualized on annual level by two figures showing the effect of array inclination ($\beta$), azimuth, deviation from south ($\gamma$) and shading of the horizon ($\nu$). In the figure depicting the shading effect, the array is due south.

It should be noted, that the conversion coefficients depend, in addition to latitude, on the local climate and are subject to some variations in the calculation models used. Additionally, the nature of the ground cover has an impact on the radiation on inclined surfaces. In the calculations a constant ground reflectance (albedo) of 0.20 was assumed, typical for most natural ground surfaces. Snow cover or other strongly reflective materials may significantly increase the insolation on steeply inclined and vertical surfaces. However, the effect is usually insignificant on an annual level. The horizontal shading is given in degrees from the horizontal and assumed to be uniform over all azimuth angles. In addition to unobstructed horizon (0° shading), two shading angles are given: 20° representing moderate shading and 40° for severe shading. Objects in the vicinity of the array that cast a sharp shadow on part of it may have a more dramatic impact on the array output. This effect is discussed in Chapter 6 of this book.

The insolation given refers to long-term averages. However, depending on climate the insolation varies from year to year. The relative variation can be especially pronounced in high latitude locations during the low insolation months. Also, sharp geographical changes may result in markedly different insolation levels for neighbouring locations e.g. in coastal or mountaineous areas. Due to the number of factors affecting the available insolation on the PV-array, primary input for calculations should be local weather data, and the following information is intended as a rough guideline only.

Table I.1: monthly and annual averages of daily insolation (kWh/m², day) for selected locations. The three temperatures indicated are minimum and maximum of monthly averages as well as the annual average. Arranged by continent and decreasing latitude.

| | | Insolation (kWh/m²,day) | | | | | | | | | | | | | Temperature (°C) | | |
|---|---|---|---|---|---|---|---|---|---|---|---|---|---|---|---|---|---|
| Location | latitude(°N) | Jan | Feb | Mar | Apr | May | Jun | Jul | Aug | Sep | Oct | Nov | Dec | Year | Min | Ave | Max |
| **Europe** | | | | | | | | | | | | | | | | | |
| Tromsø, Norway | 69.7 | 0.0 | 0.3 | 1.4 | 3.1 | 4.2 | 4.6 | 4.5 | 2.9 | 1.5 | 0.5 | 0.0 | 0.0 | 1.9 | -4 | 3 | 16 |
| Sodankylä, Finland | 67.5 | 0.1 | 0.5 | 1.8 | 3.6 | 4.9 | 5.3 | 5.0 | 3.3 | 1.7 | 0.6 | 0.1 | 0.0 | 2.2 | -18 | -1 | 15 |
| Reykjavik, Iceland | 64.1 | 0.1 | 0.6 | 1.6 | 3.1 | 4.6 | 4.3 | 4.7 | 3.5 | 1.9 | 0.8 | 0.2 | 0.1 | 2.1 | 0 | 5 | 11 |
| Östersund, Sweden | 63.2 | 0.2 | 0.9 | 2.4 | 3.9 | 5.1 | 5.6 | 5.2 | 4.1 | 2.2 | 0.9 | 0.3 | 0.1 | 2.6 | -8 | 3 | 14 |
| Bergen, Norway | 60.4 | 0.2 | 0.6 | 1.7 | 3.4 | 3.9 | 5.0 | 4.8 | 3.2 | 1.7 | 0.8 | 0.3 | 0.1 | 2.2 | 1 | 8 | 15 |
| Helsinki, Finland | 60.3 | 0.3 | 0.8 | 2.2 | 3.4 | 5.1 | 6.0 | 5.2 | 4.1 | 2.3 | 1.0 | 0.3 | 0.2 | 2.6 | -10 | 4 | 16 |
| Stockholm, Sweden | 59.3 | 0.3 | 1.0 | 2.2 | 3.6 | 5.2 | 5.9 | 5.2 | 4.1 | 2.6 | 1.2 | 0.5 | 0.2 | 2.7 | -3 | 7 | 16 |
| Edinburgh, The United Kingdom | 56.0 | 0.4 | 1.1 | 2.1 | 3.3 | 4.0 | 4.6 | 4.2 | 3.5 | 2.5 | 1.4 | 0.7 | 0.3 | 2.4 | 3 | 9 | 15 |
| Moscow, Russia | 55.8 | 0.5 | 0.9 | 2.6 | 3.1 | 4.7 | 5.4 | 5.5 | 4.3 | 2.3 | 1.1 | 0.6 | 0.4 | 2.6 | -10 | 4 | 18 |
| Copenhagen, Denmark | 55.7 | 0.4 | 1.1 | 2.3 | 3.9 | 5.0 | 6.1 | 5.3 | 4.4 | 2.8 | 1.5 | 0.6 | 0.3 | 2.8 | -1 | 8 | 17 |
| Hamburg, Germany | 53.6 | 0.5 | 1.1 | 2.2 | 3.6 | 4.7 | 5.4 | 4.8 | 4.3 | 2.8 | 1.5 | 0.7 | 0.4 | 2.7 | 1 | 9 | 17 |
| Dublin, Ireland | 53.4 | 0.7 | 1.5 | 2.7 | 4.2 | 5.2 | 6.1 | 5.3 | 4.4 | 3.3 | 1.9 | 1.0 | 0.6 | 3.1 | 4 | 10 | 15 |
| Berlin, Germany | 52.5 | 0.6 | 1.1 | 2.4 | 3.5 | 4.8 | 5.4 | 5.3 | 4.6 | 3.0 | 1.6 | 0.8 | 0.5 | 2.8 | -1 | 9 | 18 |
| Warsaw, Poland | 52.3 | 0.5 | 1.0 | 2.3 | 3.3 | 4.6 | 5.4 | 5.0 | 4.6 | 2.9 | 1.4 | 0.6 | 0.4 | 2.7 | -4 | 8 | 19 |
| Amsterdam, The Netherlands | 52.1 | 0.6 | 1.3 | 2.2 | 3.6 | 4.7 | 5.2 | 4.6 | 4.3 | 2.9 | 1.7 | 0.8 | 0.5 | 2.7 | 2 | 10 | 17 |
| London, The United Kingdom | 51.5 | 0.5 | 1.1 | 2.1 | 3.0 | 4.1 | 5.0 | 4.4 | 3.6 | 2.7 | 1.6 | 0.8 | 0.5 | 2.5 | 4 | 11 | 18 |
| Kiev, Ukraine | 50.4 | 0.8 | 1.5 | 2.6 | 3.3 | 5.6 | 5.6 | 5.7 | 4.8 | 3.3 | 2.1 | 0.7 | 0.6 | 3.1 | -6 | 8 | 20 |
| Frankfurt, Germany | 50.0 | 0.7 | 1.5 | 2.6 | 4.0 | 5.0 | 5.3 | 5.3 | 4.5 | 3.4 | 1.8 | 0.9 | 0.6 | 3.0 | 0 | 10 | 19 |
| Paris, France | 48.8 | 0.8 | 1.6 | 2.7 | 4.0 | 4.8 | 5.6 | 5.6 | 4.6 | 3.5 | 2.1 | 1.0 | 0.7 | 3.1 | 3 | 11 | 19 |
| Vienna, Austria | 48.2 | 0.8 | 1.4 | 2.6 | 4.0 | 5.1 | 5.3 | 5.4 | 4.5 | 3.3 | 2.0 | 1.0 | 0.7 | 3.0 | -1 | 10 | 20 |
| Zürich, Switzerland | 47.4 | 0.8 | 1.6 | 2.7 | 3.9 | 5.0 | 5.5 | 5.8 | 4.6 | 3.6 | 2.0 | 1.0 | 0.7 | 3.1 | -1 | 9 | 18 |
| Innsbruck, Austria | 47.3 | 1.3 | 2.1 | 3.4 | 4.5 | 5.3 | 5.4 | 5.4 | 4.7 | 4.0 | 2.6 | 1.4 | 1.1 | 3.4 | -2 | 9 | 18 |
| Clermont-Ferrand, France | 45.7 | 1.2 | 1.9 | 2.9 | 4.1 | 4.9 | 5.6 | 6.0 | 4.8 | 3.9 | 2.5 | 1.4 | 0.9 | 3.3 | 3 | 11 | 20 |
| Venice, Italy | 45.5 | 1.1 | 1.7 | 3.3 | 4.4 | 5.1 | 5.9 | 6.2 | 5.5 | 4.1 | 2.6 | 1.4 | 1.1 | 3.5 | 3 | 14 | 23 |
| Bordeaux, France | 44.8 | 1.3 | 2.1 | 3.5 | 4.7 | 5.5 | 6.1 | 6.4 | 5.1 | 4.1 | 2.9 | 1.5 | 1.0 | 3.7 | 5 | 12 | 20 |
| Bucharest, Romania | 44.5 | 1.3 | 2.1 | 3.5 | 4.7 | 6.1 | 6.6 | 6.4 | 5.7 | 4.3 | 2.9 | 1.5 | 1.1 | 3.9 | -2 | 11 | 22 |
| Nice, France | 43.6 | 1.7 | 2.5 | 3.9 | 5.3 | 6.1 | 6.8 | 7.1 | 5.9 | 4.6 | 3.3 | 2.0 | 1.6 | 4.2 | 10 | 16 | 24 |
| Oviedo, Spain | 43.1 | 1.7 | 2.3 | 3.1 | 4.0 | 4.9 | 4.8 | 4.8 | 4.3 | 3.8 | 2.8 | 2.0 | 1.4 | 3.3 | 8 | 13 | 19 |
| Rome, Italy | 41.8 | 1.7 | 2.5 | 3.8 | 5.0 | 6.0 | 6.6 | 6.7 | 6.2 | 4.7 | 3.3 | 2.0 | 1.5 | 4.2 | 7 | 16 | 25 |
| Madrid, Spain | 40.4 | 1.7 | 2.6 | 4.2 | 5.4 | 6.2 | 6.7 | 7.2 | 6.5 | 4.8 | 3.2 | 2.0 | 1.8 | 4.4 | 5 | 14 | 24 |
| Ankara, Turkey | 40.0 | 1.8 | 2.5 | 3.9 | 5.3 | 6.5 | 7.5 | 7.8 | 7.0 | 5.5 | 3.8 | 2.4 | 1.5 | 4.6 | 0 | 12 | 22 |
| Lisbon, Portugal | 38.7 | 2.0 | 3.0 | 4.3 | 5.5 | 6.7 | 7.2 | 7.5 | 7.0 | 5.2 | 3.7 | 2.5 | 2.2 | 4.7 | 11 | 17 | 23 |
| Athens, Greece | 38.0 | 1.8 | 2.6 | 3.8 | 5.1 | 6.4 | 6.8 | 6.9 | 6.2 | 4.9 | 3.4 | 2.3 | 1.7 | 4.3 | 9 | 18 | 28 |
| Trapani, Italy | 37.9 | 2.3 | 3.0 | 4.3 | 5.6 | 6.7 | 7.0 | 7.4 | 7.0 | 5.4 | 3.9 | 2.7 | 2.1 | 4.8 | 12 | 18 | 25 |
| Almeria, Spain | 36.8 | 2.7 | 3.5 | 4.3 | 5.5 | 6.7 | 7.2 | 7.4 | 6.8 | 5.3 | 4.0 | 2.9 | 2.5 | 5.0 | 12 | 18 | 25 |

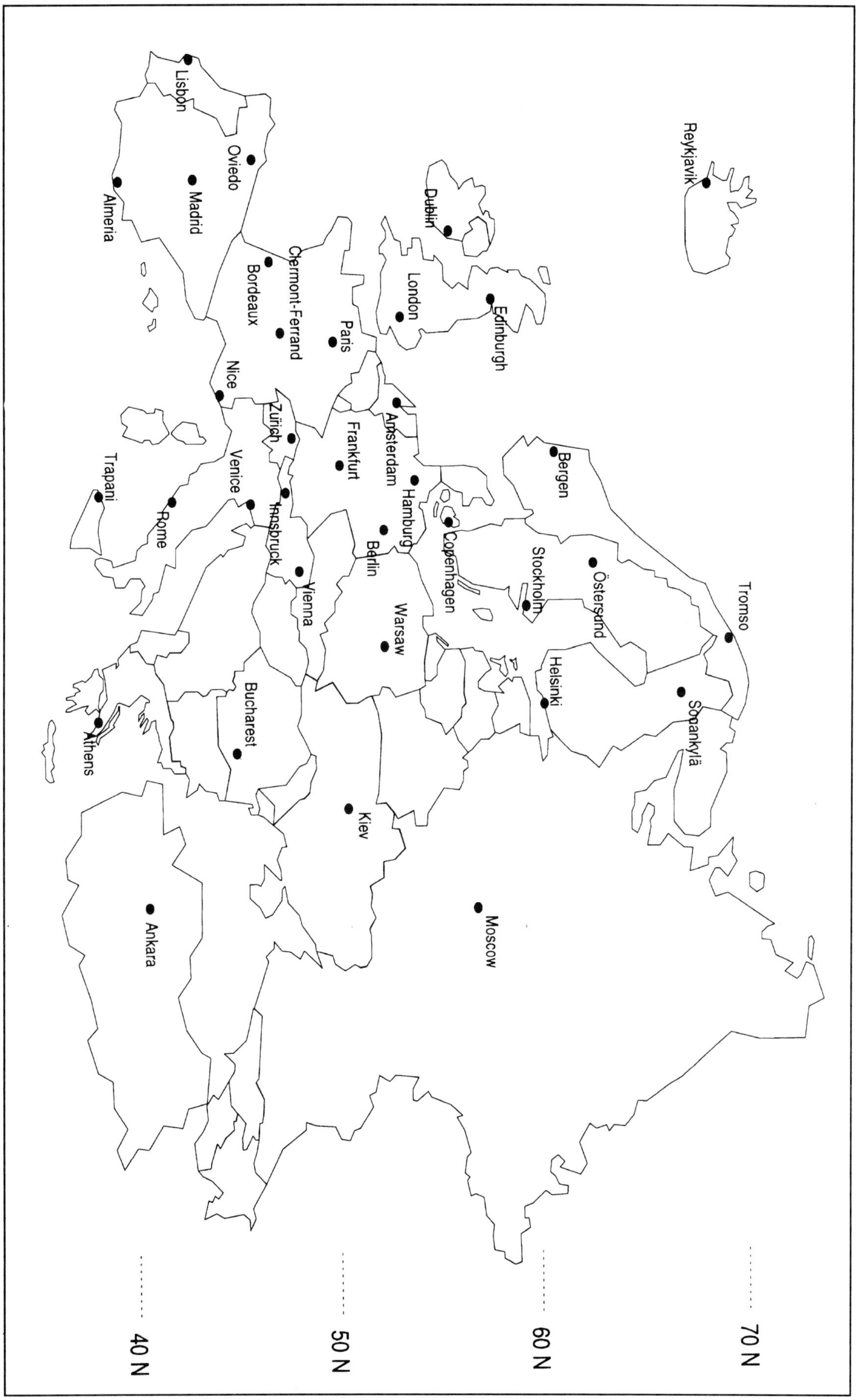
Lisbon
Oviedo
Madrid
Almeria
Reykjavik
Dublin
London
Edinburgh
Bordeaux
Clermont-Ferrand
Paris
Nice
Zürich
Amsterdam
Frankfurt
Bergen
Trapani
Venice
Hamburg
Rome
Innsbruck
Berlin
Copenhagen
Stockholm
Östersund
Vienna
Warsaw
Tromso
Helsinki
Sodankylä
Bucharest
Athens
Kiev
Ankara
Moscow
40 N
50 N
60 N
70 N

**North America**

| | | | | | | | | | | | | | | | | | |
|---|---|---|---|---|---|---|---|---|---|---|---|---|---|---|---|---|---|
| Resolute (Northwest Terr.), Canada | 74.7 | 0.0 | 0.2 | 1.5 | 4.1 | 6.4 | 6.9 | 5.2 | 3.1 | 1.4 | 0.4 | 0.0 | 0.0 | 2.4 | -33 | -17 | 4 |
| Norman Wells (Northwest Terr.), Canada | 65.3 | 0.1 | 0.8 | 2.3 | 4.3 | 5.6 | 6.4 | 5.6 | 4.0 | 2.3 | 0.8 | 0.2 | 0.0 | 2.7 | -29 | -6 | 16 |
| Fairbanks (Alaska), USA | 64.8 | 0.1 | 0.7 | 2.1 | 3.8 | 5.1 | 5.5 | 4.9 | 3.5 | 2.2 | 0.9 | 0.2 | 0.0 | 2.4 | -24 | -4 | 16 |
| Baker Lake (Northwest Terr.), Canada | 64.3 | 0.2 | 0.9 | 2.9 | 4.9 | 6.2 | 6.0 | 5.4 | 3.8 | 2.1 | 1.0 | 0.3 | 0.1 | 2.8 | -33 | -12 | 11 |
| White Horse (Yukon), Canada | 60.7 | 0.4 | 1.1 | 2.6 | 4.4 | 5.4 | 5.8 | 5.2 | 4.2 | 2.5 | 1.2 | 0.4 | 0.2 | 2.8 | -21 | -1 | 12 |
| Churchill (Manitoba), Canada | 58.8 | 0.8 | 1.6 | 3.4 | 5.2 | 5.6 | 6.1 | 5.7 | 4.2 | 2.4 | 1.2 | 0.7 | 0.4 | 3.1 | -28 | -7 | 12 |
| Prince George (Brit. Columbia), Canada | 53.9 | 0.8 | 1.5 | 2.8 | 4.4 | 5.0 | 6.0 | 5.8 | 4.7 | 3.1 | 1.7 | 0.9 | 0.5 | 3.1 | -12 | 3 | 15 |
| Big Trout Lake (Ontario), Canada | 53.8 | 1.1 | 2.1 | 3.6 | 5.3 | 5.6 | 5.4 | 5.7 | 4.3 | 2.7 | 1.6 | 0.9 | 0.8 | 3.3 | -25 | -3 | 16 |
| Edmonton (Alberta), Canada | 53.6 | 1.0 | 2.0 | 3.5 | 4.9 | 5.8 | 6.3 | 6.4 | 5.0 | 3.6 | 2.2 | 1.2 | 0.8 | 3.6 | -15 | 3 | 17 |
| Nitchequon (Quebec), Canada | 53.2 | 1.0 | 2.2 | 3.6 | 5.2 | 5.5 | 5.2 | 4.7 | 4.0 | 2.6 | 1.5 | 0.9 | 0.8 | 3.1 | -23 | -4 | 14 |
| Winnipeg (Manitoba), Canada | 49.9 | 1.5 | 2.5 | 3.9 | 4.9 | 5.8 | 6.3 | 6.4 | 5.3 | 3.7 | 2.3 | 1.3 | 1.1 | 3.7 | -19 | 2 | 20 |
| Vancouver (Brit. Columbia), Canada | 49.3 | 0.8 | 1.5 | 2.8 | 4.2 | 5.6 | 6.1 | 6.4 | 5.2 | 3.7 | 2.1 | 1.0 | 0.6 | 3.3 | 3 | 10 | 17 |
| St John's (Newfoundland), Canada | 47.5 | 1.2 | 2.0 | 2.9 | 3.8 | 4.6 | 5.4 | 5.5 | 4.4 | 3.3 | 1.9 | 1.1 | 0.8 | 3.1 | -4 | 5 | 16 |
| Billings (Montana), USA | 45.8 | 1.5 | 2.4 | 3.8 | 4.8 | 6.0 | 6.9 | 7.5 | 6.4 | 4.6 | 3.1 | 1.8 | 1.3 | 4.2 | -6 | 8 | 22 |
| Montreal (Quebec), Canada | 45.5 | 1.5 | 2.4 | 3.5 | 4.4 | 5.3 | 5.6 | 5.8 | 4.8 | 3.7 | 2.2 | 1.3 | 1.1 | 3.5 | -10 | 6 | 21 |
| Pierre (South Dakota), USA | 44.4 | 1.7 | 2.5 | 3.8 | 5.1 | 6.2 | 6.9 | 7.2 | 6.3 | 4.7 | 3.3 | 2.0 | 1.4 | 4.3 | -9 | 8 | 24 |
| Medford (Oregon), USA | 42.4 | 1.3 | 2.3 | 3.6 | 5.2 | 6.4 | 7.2 | 7.8 | 6.7 | 5.0 | 3.1 | 1.6 | 1.1 | 4.3 | 3 | 12 | 22 |
| Chigago (Illinois), USA | 41.8 | 1.6 | 2.4 | 3.5 | 4.6 | 5.6 | 6.3 | 6.1 | 5.4 | 4.3 | 3.1 | 1.9 | 1.3 | 3.8 | -4 | 10 | 24 |
| New York (New York), USA | 40.8 | 1.6 | 2.3 | 3.3 | 4.3 | 5.2 | 5.4 | 5.3 | 4.7 | 3.8 | 2.8 | 1.7 | 1.3 | 3.5 | 0 | 13 | 25 |
| Salt Lake City (Utah), USA | 40.8 | 2.0 | 3.1 | 4.6 | 6.0 | 7.4 | 8.0 | 8.2 | 7.1 | 5.8 | 4.1 | 2.5 | 1.8 | 5.0 | -2 | 10 | 25 |
| Pittsburg (Pennsylvania), USA | 40.5 | 1.3 | 2.0 | 3.0 | 4.2 | 5.0 | 5.6 | 5.3 | 4.8 | 3.8 | 2.8 | 1.6 | 1.1 | 3.4 | -2 | 10 | 22 |
| Denver (Colorado), USA | 39.8 | 2.6 | 3.6 | 4.8 | 5.9 | 6.7 | 7.4 | 7.2 | 6.4 | 5.4 | 4.1 | 2.8 | 2.3 | 4.9 | -1 | 10 | 23 |
| Kansas City (Missouri), USA | 39.3 | 2.0 | 2.8 | 3.8 | 5.0 | 5.9 | 6.6 | 6.6 | 5.9 | 4.6 | 3.4 | 2.3 | 1.8 | 4.2 | -3 | 10 | 25 |
| San Francisco (California), USA | 37.6 | 2.2 | 3.2 | 4.6 | 6.1 | 7.0 | 7.5 | 7.5 | 6.7 | 5.5 | 3.9 | 2.6 | 2.0 | 4.9 | 9 | 14 | 18 |
| Norfolk (Virginia), USA | 36.9 | 2.1 | 2.9 | 4.0 | 5.3 | 6.0 | 6.3 | 5.8 | 5.3 | 4.4 | 3.4 | 2.6 | 2.0 | 4.2 | 5 | 15 | 25 |
| Nashville (Tennesee), USA | 36.1 | 1.8 | 2.6 | 3.6 | 4.9 | 5.8 | 6.2 | 6.0 | 5.5 | 4.4 | 3.5 | 2.2 | 1.6 | 4.0 | 3 | 15 | 26 |
| Albuquerque (New Mexico), USA | 35.1 | 3.2 | 4.2 | 5.6 | 7.0 | 8.0 | 8.4 | 7.8 | 7.2 | 6.2 | 4.9 | 3.6 | 2.9 | 5.8 | 2 | 13 | 26 |
| Los Angeles (California), USA | 33.9 | 2.9 | 3.8 | 5.1 | 6.2 | 6.5 | 6.7 | 7.3 | 6.6 | 5.3 | 4.2 | 3.3 | 2.7 | 5.0 | 13 | 17 | 21 |
| Phoenix (Arizona), USA | 33.4 | 3.2 | 4.3 | 5.7 | 7.4 | 8.4 | 8.6 | 7.8 | 7.2 | 6.4 | 5.0 | 3.6 | 2.9 | 5.9 | 11 | 19 | 31 |
| Charleston (South Carolina), USA | 32.9 | 2.3 | 3.1 | 4.2 | 5.5 | 5.9 | 5.8 | 5.7 | 5.0 | 4.4 | 3.8 | 2.9 | 2.3 | 4.2 | 9 | 18 | 27 |
| Dallas (Texas), USA | 32.9 | 2.6 | 3.4 | 4.5 | 5.1 | 6.0 | 6.7 | 6.7 | 6.2 | 5.0 | 4.0 | 3.0 | 2.5 | 4.6 | 7 | 19 | 30 |
| New Orleans (Louisiana), USA | 30.0 | 2.6 | 3.5 | 4.5 | 5.6 | 6.2 | 6.3 | 5.7 | 5.4 | 4.4 | 4.2 | 3.1 | 2.5 | 4.5 | 12 | 20 | 28 |
| Miami (Florida), USA | 25.8 | 3.3 | 4.1 | 5.1 | 5.9 | 5.8 | 5.4 | 5.6 | 5.1 | 4.6 | 4.1 | 3.5 | 3.2 | 4.6 | 20 | 24 | 28 |
| Mexico City, Mexico | 19.4 | 4.6 | 5.1 | 5.9 | 6.1 | 5.6 | 5.5 | 5.1 | 4.9 | 4.4 | 4.6 | 4.4 | 4.3 | 5.0 | 12 | 16 | 19 |

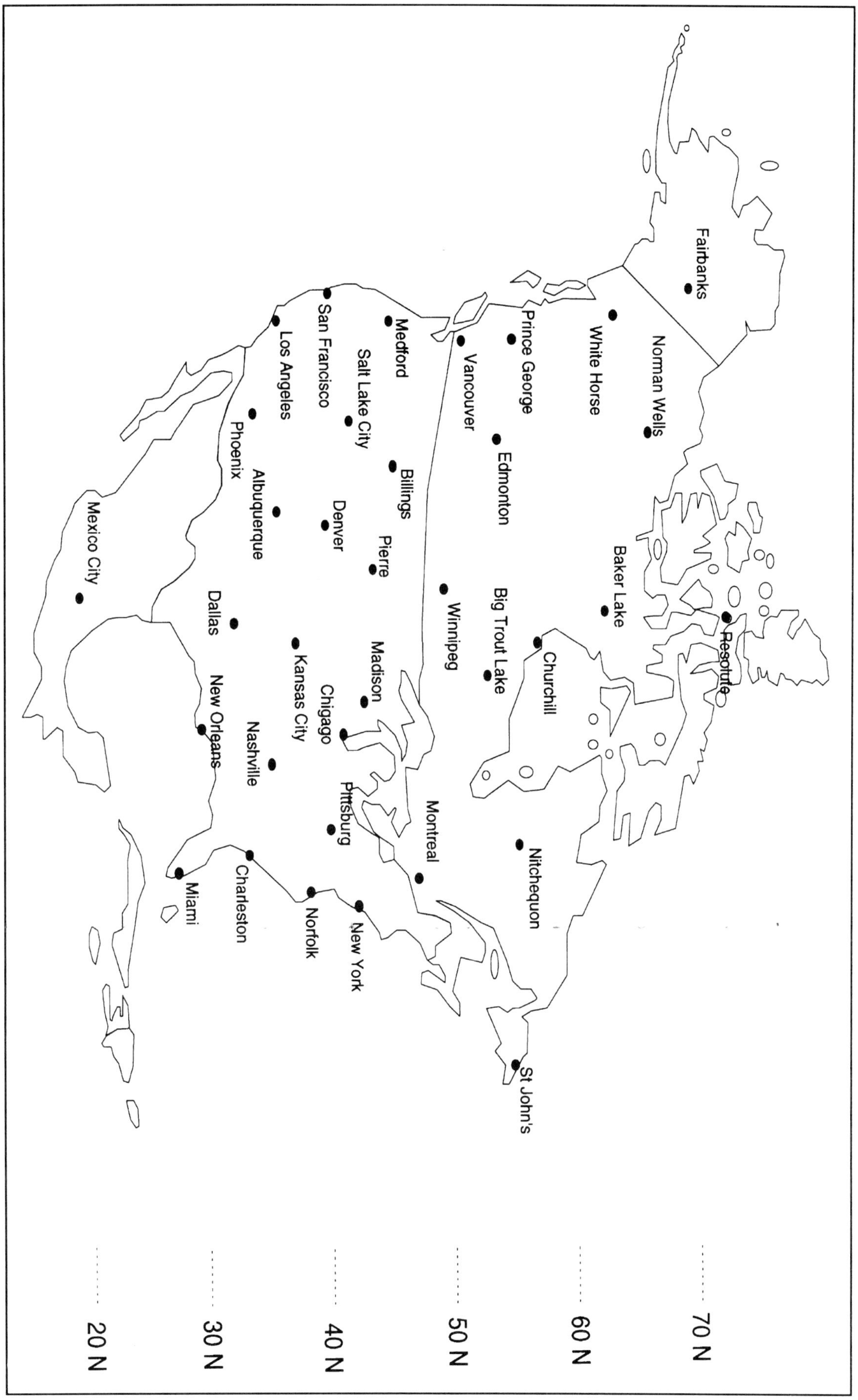
Fairbanks
Norman Wells
White Horse
Prince George
Vancouver
Medford
San Francisco
Salt Lake City
Los Angeles
Phoenix
Edmonton
Billings
Albuquerque
Denver
Pierre
Mexico City
Baker Lake
Big Trout Lake
Winnipeg
Resolute
Churchill
Dallas
Madison
Kansas City
New Orleans
Chigago
Nashville
Pittsburg
Montreal
Nitchequon
Charleston
Miami
Norfolk
New York
St John's
70 N
60 N
50 N
40 N
30 N
20 N

**Pacific**

| | | | | | | | | | | | | | | | | |
|---|---|---|---|---|---|---|---|---|---|---|---|---|---|---|---|---|
| Sapporo, Japan | 43.0 | 1.6 | 2.3 | 3.3 | 4.2 | 4.7 | 4.8 | 4.5 | 4.1 | 3.5 | 2.6 | 1.7 | 1.3 | 3.2 | -5 | 8 | 22 |
| Tokyo, Japan | 35.7 | 2.2 | 2.6 | 3.1 | 3.6 | 3.9 | 3.5 | 3.8 | 3.8 | 2.9 | 2.3 | 2.1 | 1.9 | 3.0 | 4 | 15 | 26 |
| Kagoshima, Japan | 31.6 | 2.4 | 2.8 | 3.6 | 3.9 | 4.2 | 4.1 | 4.7 | 5.0 | 3.9 | 3.4 | 2.6 | 2.3 | 3.6 | 7 | 17 | 27 |
| Honolulu (Hawaii), USA | 21.3 | 3.3 | 4.4 | 5.1 | 5.7 | 6.1 | 6.3 | 6.3 | 6.2 | 5.7 | 4.9 | 4.0 | 3.6 | 5.2 | 22 | 25 | 27 |
| Darwin, Australia | -12.4 | 5.1 | 5.3 | 5.6 | 5.1 | 5.2 | 5.1 | 5.3 | 6.1 | 6.4 | 6.5 | 6.2 | 5.6 | 5.6 | 25 | 27 | 29 |
| Alice Springs, Australia | -23.6 | 7.5 | 7.2 | 6.5 | 5.4 | 5.4 | 3.9 | 4.2 | 5.3 | 6.4 | 6.9 | 7.2 | 7.4 | 6.0 | 12 | 21 | 28 |
| Perth, Australia | -31.9 | 7.0 | 6.7 | 5.4 | 3.9 | 2.9 | 2.5 | 2.8 | 3.7 | 5.0 | 6.0 | 6.5 | 7.2 | 5.0 | 13 | 16 | 25 |
| Sydney, Australia | -33.8 | 6.2 | 5.3 | 5.1 | 3.7 | 3.0 | 2.5 | 2.9 | 3.7 | 4.6 | 6.0 | 6.8 | 6.4 | 4.7 | 12 | 17 | 22 |
| Wellington, New Zealand | -41.3 | 6.2 | 5.4 | 4.0 | 2.8 | 1.7 | 1.4 | 1.5 | 2.1 | 3.4 | 4.8 | 5.9 | 6.2 | 3.8 | 8 | 13 | 17 |

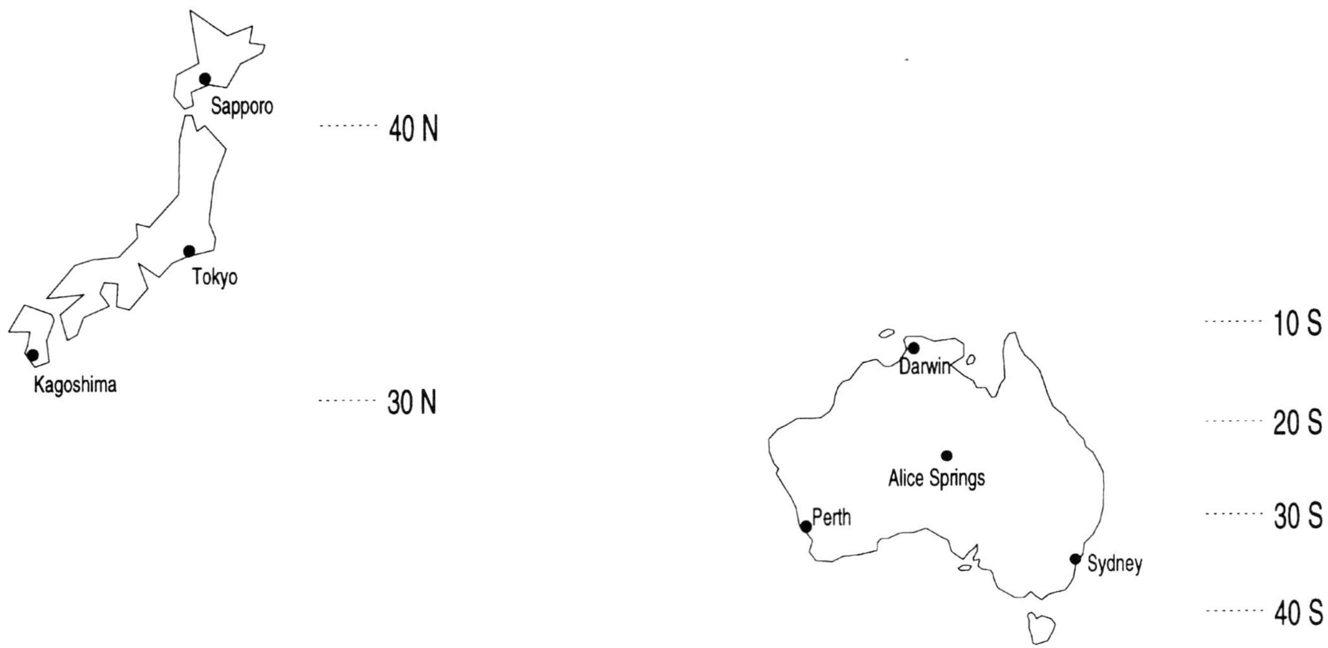

Phoenix (Arizona), USA 33.4o N

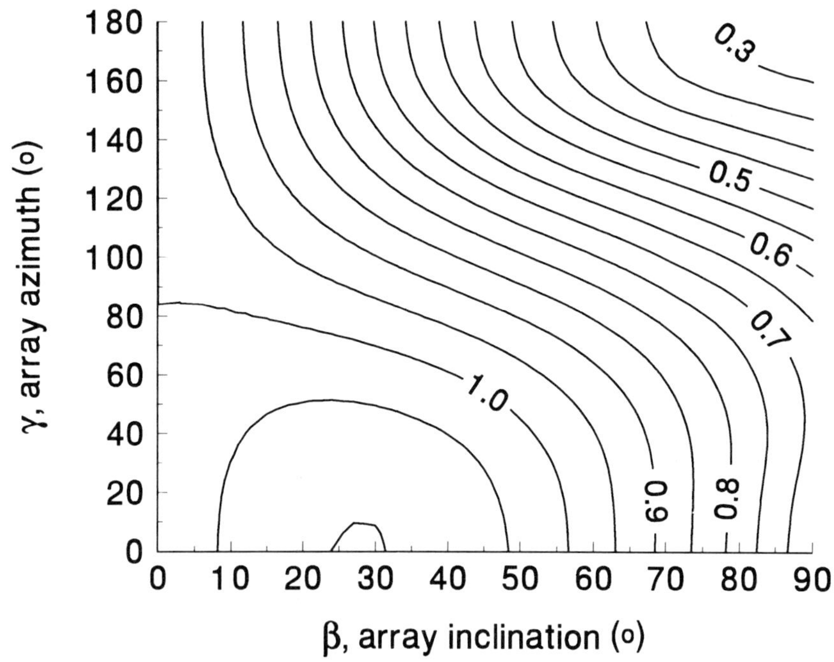


Phoenix (Arizona), USA 33.4 oN

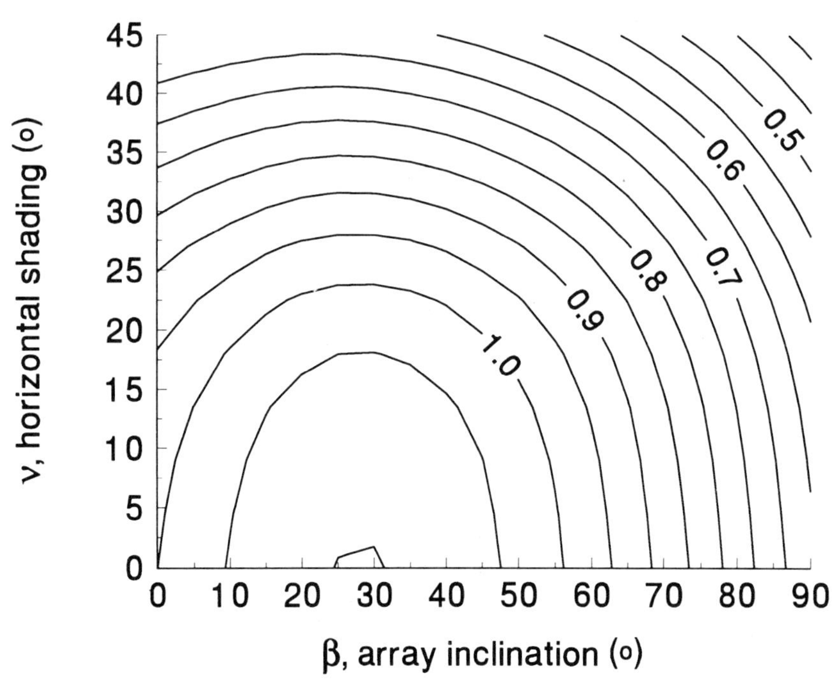

Phoenix (Arizona), USA 33.4°N

| β | γ | ν | Jan | Feb | Mar | Apr | May | Jun | Jul | Aug | Sep | Oct | Nov | Dec | Annual |
|---|---|---|---|---|---|---|---|---|---|---|---|---|---|---|---|
| tracking | | 0 | 2.0 | 1.8 | 1.6 | 1.5 | 1.5 | 1.4 | 1.4 | 1.5 | 1.6 | 1.7 | 2.0 | 2.0 | 1.6 |
| tracking | | 20 | 1.6 | 1.5 | 1.4 | 1.3 | 1.3 | 1.3 | 1.3 | 1.3 | 1.3 | 1.4 | 1.5 | 1.5 | 1.3 |
| tracking | | 40 | 0.4 | 0.7 | 0.9 | 1.0 | 0.9 | 0.9 | 0.9 | 1.0 | 0.9 | 0.9 | 0.5 | 0.5 | 0.9 |
| 0 | 0 | 0 | 1.0 | 1.0 | 1.0 | 1.0 | 1.0 | 1.0 | 1.0 | 1.0 | 1.0 | 1.0 | 1.0 | 1.0 | 1.0 |
| 0 | 0 | 20 | 0.9 | 0.9 | 1.0 | 0.9 | 1.0 | 1.0 | 1.0 | 1.0 | 1.0 | 0.9 | 0.9 | 0.9 | 0.9 |
| 0 | 0 | 40 | 0.3 | 0.5 | 0.7 | 0.8 | 0.8 | 0.8 | 0.8 | 0.8 | 0.8 | 0.6 | 0.3 | 0.4 | 0.7 |
| 15 | 0 | 0 | 1.3 | 1.2 | 1.1 | 1.0 | 1.0 | 1.0 | 1.0 | 1.0 | 1.1 | 1.2 | 1.3 | 1.3 | 1.1 |
| 15 | 0 | 20 | 1.1 | 1.1 | 1.1 | 1.0 | 1.0 | 0.9 | 1.0 | 1.0 | 1.0 | 1.1 | 1.1 | 1.1 | 1.0 |
| 15 | 0 | 40 | 0.3 | 0.6 | 0.8 | 0.9 | 0.8 | 0.8 | 0.8 | 0.9 | 0.8 | 0.7 | 0.4 | 0.4 | 0.7 |
| 15 | 45 | 0 | 1.2 | 1.1 | 1.1 | 1.0 | 1.0 | 1.0 | 1.0 | 1.0 | 1.1 | 1.1 | 1.2 | 1.2 | 1.1 |
| 15 | 45 | 20 | 1.1 | 1.0 | 1.0 | 1.0 | 1.0 | 0.9 | 1.0 | 1.0 | 1.0 | 1.0 | 1.0 | 1.0 | 1.0 |
| 15 | 45 | 40 | 0.3 | 0.6 | 0.8 | 0.8 | 0.8 | 0.8 | 0.8 | 0.8 | 0.8 | 0.7 | 0.4 | 0.4 | 0.7 |
| 15 | 90 | 0 | 1.0 | 1.0 | 1.0 | 1.0 | 1.0 | 1.0 | 1.0 | 1.0 | 1.0 | 1.0 | 1.0 | 1.0 | 1.0 |
| 15 | 90 | 20 | 0.9 | 0.9 | 0.9 | 0.9 | 0.9 | 0.9 | 0.9 | 0.9 | 0.9 | 0.9 | 0.9 | 0.8 | 0.9 |
| 15 | 90 | 40 | 0.3 | 0.5 | 0.7 | 0.8 | 0.8 | 0.8 | 0.8 | 0.8 | 0.7 | 0.6 | 0.3 | 0.4 | 0.7 |
| 15 | 135 | 0 | 0.8 | 0.8 | 0.9 | 0.9 | 1.0 | 1.0 | 1.0 | 0.9 | 0.9 | 0.8 | 0.8 | 0.8 | 0.9 |
| 15 | 135 | 20 | 0.7 | 0.8 | 0.8 | 0.9 | 0.9 | 0.9 | 0.9 | 0.9 | 0.9 | 0.8 | 0.7 | 0.7 | 0.9 |
| 15 | 135 | 40 | 0.3 | 0.5 | 0.6 | 0.7 | 0.7 | 0.7 | 0.8 | 0.8 | 0.7 | 0.6 | 0.3 | 0.3 | 0.6 |
| 15 | 180 | 0 | 0.7 | 0.7 | 0.8 | 0.9 | 1.0 | 1.0 | 1.0 | 0.9 | 0.9 | 0.8 | 0.7 | 0.7 | 0.9 |
| 15 | 180 | 20 | 0.6 | 0.7 | 0.8 | 0.8 | 0.9 | 0.9 | 0.9 | 0.9 | 0.8 | 0.7 | 0.6 | 0.6 | 0.8 |
| 15 | 180 | 40 | 0.3 | 0.4 | 0.6 | 0.7 | 0.7 | 0.7 | 0.7 | 0.7 | 0.6 | 0.5 | 0.3 | 0.3 | 0.6 |
| 30 | 0 | 0 | 1.5 | 1.4 | 1.2 | 1.0 | 0.9 | 0.9 | 0.9 | 1.0 | 1.1 | 1.3 | 1.5 | 1.5 | 1.1 |
| 30 | 0 | 20 | 1.3 | 1.2 | 1.1 | 1.0 | 0.9 | 0.9 | 0.9 | 1.0 | 1.1 | 1.2 | 1.3 | 1.3 | 1.0 |
| 30 | 0 | 40 | 0.4 | 0.6 | 0.8 | 0.9 | 0.8 | 0.7 | 0.8 | 0.8 | 0.8 | 0.8 | 0.4 | 0.4 | 0.7 |
| 30 | 45 | 0 | 1.3 | 1.2 | 1.1 | 1.0 | 0.9 | 0.9 | 0.9 | 1.0 | 1.1 | 1.2 | 1.3 | 1.3 | 1.1 |
| 30 | 45 | 20 | 1.2 | 1.1 | 1.1 | 0.9 | 0.9 | 0.9 | 0.9 | 0.9 | 1.0 | 1.1 | 1.1 | 1.1 | 1.0 |
| 30 | 45 | 40 | 0.3 | 0.6 | 0.8 | 0.8 | 0.8 | 0.7 | 0.7 | 0.8 | 0.8 | 0.7 | 0.4 | 0.4 | 0.7 |
| 30 | 90 | 0 | 1.0 | 0.9 | 0.9 | 0.9 | 0.9 | 0.9 | 0.9 | 1.0 | 0.9 | 0.9 | 1.0 | 0.9 | 0.9 |
| 30 | 90 | 20 | 0.8 | 0.8 | 0.9 | 0.8 | 0.9 | 0.9 | 0.9 | 0.9 | 0.9 | 0.8 | 0.8 | 0.8 | 0.9 |
| 30 | 90 | 40 | 0.3 | 0.5 | 0.6 | 0.7 | 0.7 | 0.7 | 0.7 | 0.7 | 0.7 | 0.6 | 0.3 | 0.4 | 0.6 |
| 30 | 135 | 0 | 0.6 | 0.6 | 0.7 | 0.8 | 0.9 | 0.9 | 0.9 | 0.9 | 0.8 | 0.7 | 0.6 | 0.6 | 0.8 |
| 30 | 135 | 20 | 0.5 | 0.6 | 0.7 | 0.7 | 0.8 | 0.9 | 0.8 | 0.8 | 0.7 | 0.6 | 0.5 | 0.5 | 0.7 |
| 30 | 135 | 40 | 0.3 | 0.4 | 0.5 | 0.6 | 0.7 | 0.7 | 0.7 | 0.7 | 0.6 | 0.4 | 0.3 | 0.3 | 0.6 |
| 30 | 180 | 0 | 0.4 | 0.5 | 0.6 | 0.7 | 0.9 | 0.9 | 0.9 | 0.8 | 0.7 | 0.5 | 0.4 | 0.4 | 0.7 |
| 30 | 180 | 20 | 0.4 | 0.5 | 0.6 | 0.7 | 0.8 | 0.8 | 0.8 | 0.7 | 0.6 | 0.5 | 0.4 | 0.4 | 0.7 |
| 30 | 180 | 40 | 0.3 | 0.3 | 0.5 | 0.6 | 0.6 | 0.7 | 0.7 | 0.6 | 0.5 | 0.4 | 0.3 | 0.3 | 0.5 |
| 45 | 0 | 0 | 1.6 | 1.4 | 1.2 | 1.0 | 0.8 | 0.8 | 0.8 | 0.9 | 1.1 | 1.3 | 1.6 | 1.7 | 1.1 |
| 45 | 0 | 20 | 1.4 | 1.3 | 1.1 | 0.9 | 0.8 | 0.8 | 0.8 | 0.9 | 1.0 | 1.2 | 1.4 | 1.4 | 1.0 |
| 45 | 0 | 40 | 0.4 | 0.7 | 0.8 | 0.8 | 0.7 | 0.7 | 0.7 | 0.8 | 0.8 | 0.8 | 0.4 | 0.4 | 0.7 |
| 45 | 45 | 0 | 1.4 | 1.3 | 1.1 | 1.0 | 0.9 | 0.8 | 0.8 | 0.9 | 1.0 | 1.2 | 1.4 | 1.4 | 1.0 |
| 45 | 45 | 20 | 1.2 | 1.1 | 1.0 | 0.9 | 0.8 | 0.8 | 0.8 | 0.9 | 1.0 | 1.1 | 1.2 | 1.1 | 0.9 |
| 45 | 45 | 40 | 0.4 | 0.6 | 0.8 | 0.8 | 0.7 | 0.6 | 0.7 | 0.8 | 0.8 | 0.7 | 0.4 | 0.4 | 0.7 |
| 45 | 90 | 0 | 0.9 | 0.9 | 0.9 | 0.9 | 0.9 | 0.9 | 0.8 | 0.9 | 0.9 | 0.9 | 0.9 | 0.9 | 0.9 |
| 45 | 90 | 20 | 0.8 | 0.8 | 0.8 | 0.8 | 0.8 | 0.8 | 0.8 | 0.8 | 0.8 | 0.8 | 0.8 | 0.7 | 0.8 |
| 45 | 90 | 40 | 0.3 | 0.5 | 0.6 | 0.6 | 0.6 | 0.6 | 0.6 | 0.7 | 0.6 | 0.5 | 0.3 | 0.4 | 0.6 |
| 45 | 135 | 0 | 0.4 | 0.5 | 0.6 | 0.7 | 0.8 | 0.8 | 0.8 | 0.7 | 0.6 | 0.5 | 0.5 | 0.4 | 0.7 |
| 45 | 135 | 20 | 0.4 | 0.4 | 0.5 | 0.6 | 0.7 | 0.7 | 0.7 | 0.7 | 0.6 | 0.4 | 0.4 | 0.4 | 0.6 |
| 45 | 135 | 40 | 0.3 | 0.3 | 0.4 | 0.5 | 0.6 | 0.6 | 0.6 | 0.5 | 0.4 | 0.3 | 0.3 | 0.3 | 0.5 |
| 45 | 180 | 0 | 0.3 | 0.3 | 0.4 | 0.6 | 0.7 | 0.8 | 0.8 | 0.6 | 0.4 | 0.3 | 0.3 | 0.3 | 0.5 |
| 45 | 180 | 20 | 0.3 | 0.3 | 0.4 | 0.5 | 0.7 | 0.7 | 0.7 | 0.6 | 0.4 | 0.3 | 0.3 | 0.3 | 0.5 |
| 45 | 180 | 40 | 0.3 | 0.3 | 0.3 | 0.4 | 0.5 | 0.5 | 0.5 | 0.5 | 0.4 | 0.3 | 0.3 | 0.3 | 0.4 |
| 60 | 0 | 0 | 1.7 | 1.4 | 1.1 | 0.8 | 0.7 | 0.6 | 0.6 | 0.8 | 1.0 | 1.3 | 1.6 | 1.7 | 1.0 |
| 60 | 0 | 20 | 1.4 | 1.3 | 1.1 | 0.8 | 0.7 | 0.6 | 0.6 | 0.8 | 1.0 | 1.2 | 1.4 | 1.4 | 0.9 |
| 60 | 0 | 40 | 0.4 | 0.6 | 0.8 | 0.7 | 0.6 | 0.5 | 0.6 | 0.7 | 0.8 | 0.8 | 0.4 | 0.4 | 0.6 |
| 60 | 45 | 0 | 1.4 | 1.2 | 1.0 | 0.9 | 0.8 | 0.7 | 0.7 | 0.8 | 1.0 | 1.1 | 1.4 | 1.4 | 0.9 |
| 60 | 45 | 20 | 1.2 | 1.1 | 0.9 | 0.8 | 0.7 | 0.7 | 0.7 | 0.8 | 0.9 | 1.0 | 1.1 | 1.1 | 0.8 |
| 60 | 45 | 40 | 0.4 | 0.6 | 0.7 | 0.7 | 0.6 | 0.5 | 0.6 | 0.7 | 0.7 | 0.7 | 0.4 | 0.4 | 0.6 |
| 60 | 90 | 0 | 0.8 | 0.8 | 0.8 | 0.8 | 0.8 | 0.8 | 0.8 | 0.8 | 0.8 | 0.8 | 0.9 | 0.8 | 0.8 |
| 60 | 90 | 20 | 0.7 | 0.7 | 0.7 | 0.7 | 0.7 | 0.7 | 0.7 | 0.7 | 0.7 | 0.7 | 0.7 | 0.6 | 0.7 |
| 60 | 90 | 40 | 0.3 | 0.4 | 0.5 | 0.5 | 0.5 | 0.5 | 0.5 | 0.6 | 0.5 | 0.5 | 0.3 | 0.4 | 0.5 |
| 60 | 135 | 0 | 0.4 | 0.4 | 0.5 | 0.6 | 0.7 | 0.7 | 0.6 | 0.6 | 0.5 | 0.4 | 0.4 | 0.4 | 0.6 |
| 60 | 135 | 20 | 0.3 | 0.4 | 0.4 | 0.5 | 0.6 | 0.6 | 0.6 | 0.5 | 0.5 | 0.3 | 0.3 | 0.3 | 0.5 |
| 60 | 135 | 40 | 0.3 | 0.3 | 0.3 | 0.4 | 0.4 | 0.4 | 0.4 | 0.4 | 0.3 | 0.3 | 0.3 | 0.3 | 0.4 |
| 60 | 180 | 0 | 0.3 | 0.3 | 0.2 | 0.4 | 0.5 | 0.6 | 0.6 | 0.5 | 0.2 | 0.2 | 0.3 | 0.3 | 0.4 |
| 60 | 180 | 20 | 0.3 | 0.3 | 0.2 | 0.3 | 0.5 | 0.6 | 0.5 | 0.4 | 0.2 | 0.2 | 0.3 | 0.3 | 0.4 |
| 60 | 180 | 40 | 0.3 | 0.3 | 0.2 | 0.3 | 0.4 | 0.4 | 0.4 | 0.3 | 0.2 | 0.2 | 0.3 | 0.3 | 0.3 |
| 75 | 0 | 0 | 1.6 | 1.3 | 1.0 | 0.7 | 0.5 | 0.4 | 0.5 | 0.6 | 0.9 | 1.2 | 1.5 | 1.7 | 0.8 |
| 75 | 0 | 20 | 1.4 | 1.2 | 0.9 | 0.7 | 0.5 | 0.4 | 0.5 | 0.6 | 0.8 | 1.1 | 1.3 | 1.3 | 0.8 |
| 75 | 0 | 40 | 0.4 | 0.6 | 0.7 | 0.6 | 0.5 | 0.4 | 0.5 | 0.6 | 0.7 | 0.7 | 0.4 | 0.4 | 0.5 |
| 75 | 45 | 0 | 1.3 | 1.1 | 0.9 | 0.7 | 0.6 | 0.6 | 0.6 | 0.7 | 0.9 | 1.0 | 1.3 | 1.3 | 0.8 |
| 75 | 45 | 20 | 1.1 | 1.0 | 0.8 | 0.6 | 0.6 | 0.5 | 0.6 | 0.7 | 0.8 | 0.9 | 1.1 | 1.0 | 0.7 |
| 75 | 45 | 40 | 0.3 | 0.5 | 0.6 | 0.6 | 0.5 | 0.4 | 0.4 | 0.5 | 0.6 | 0.6 | 0.4 | 0.4 | 0.5 |
| 75 | 90 | 0 | 0.8 | 0.8 | 0.7 | 0.7 | 0.7 | 0.7 | 0.7 | 0.7 | 0.7 | 0.7 | 0.8 | 0.8 | 0.7 |
| 75 | 90 | 20 | 0.6 | 0.6 | 0.6 | 0.6 | 0.6 | 0.6 | 0.6 | 0.6 | 0.6 | 0.6 | 0.6 | 0.5 | 0.6 |
| 75 | 90 | 40 | 0.3 | 0.4 | 0.4 | 0.5 | 0.4 | 0.4 | 0.4 | 0.5 | 0.4 | 0.4 | 0.3 | 0.3 | 0.4 |
| 75 | 135 | 0 | 0.3 | 0.4 | 0.4 | 0.5 | 0.6 | 0.6 | 0.5 | 0.5 | 0.4 | 0.4 | 0.3 | 0.4 | 0.5 |
| 75 | 135 | 20 | 0.3 | 0.3 | 0.3 | 0.4 | 0.5 | 0.5 | 0.5 | 0.4 | 0.4 | 0.3 | 0.3 | 0.3 | 0.4 |
| 75 | 135 | 40 | 0.3 | 0.3 | 0.2 | 0.3 | 0.3 | 0.3 | 0.3 | 0.3 | 0.3 | 0.2 | 0.3 | 0.3 | 0.3 |
| 75 | 180 | 0 | 0.3 | 0.3 | 0.2 | 0.2 | 0.3 | 0.4 | 0.4 | 0.3 | 0.2 | 0.2 | 0.3 | 0.3 | 0.3 |
| 75 | 180 | 20 | 0.3 | 0.3 | 0.2 | 0.2 | 0.3 | 0.4 | 0.3 | 0.3 | 0.2 | 0.2 | 0.3 | 0.3 | 0.3 |
| 75 | 180 | 40 | 0.3 | 0.3 | 0.2 | 0.2 | 0.2 | 0.3 | 0.3 | 0.2 | 0.2 | 0.2 | 0.3 | 0.3 | 0.2 |
| 90 | 0 | 0 | 1.4 | 1.1 | 0.8 | 0.5 | 0.3 | 0.3 | 0.3 | 0.4 | 0.7 | 1.0 | 1.4 | 1.5 | 0.7 |
| 90 | 0 | 20 | 1.2 | 1.0 | 0.8 | 0.5 | 0.3 | 0.3 | 0.3 | 0.4 | 0.7 | 0.9 | 1.1 | 1.2 | 0.6 |
| 90 | 0 | 40 | 0.3 | 0.5 | 0.6 | 0.5 | 0.3 | 0.3 | 0.3 | 0.4 | 0.5 | 0.6 | 0.4 | 0.4 | 0.4 |
| 90 | 45 | 0 | 1.1 | 1.0 | 0.8 | 0.6 | 0.5 | 0.5 | 0.5 | 0.6 | 0.7 | 0.9 | 1.1 | 1.2 | 0.7 |
| 90 | 45 | 20 | 1.0 | 0.8 | 0.7 | 0.5 | 0.5 | 0.4 | 0.4 | 0.5 | 0.6 | 0.7 | 0.9 | 0.9 | 0.6 |
| 90 | 45 | 40 | 0.3 | 0.5 | 0.5 | 0.4 | 0.4 | 0.3 | 0.3 | 0.4 | 0.5 | 0.5 | 0.3 | 0.4 | 0.4 |
| 90 | 90 | 0 | 0.7 | 0.7 | 0.6 | 0.6 | 0.6 | 0.6 | 0.6 | 0.6 | 0.6 | 0.6 | 0.7 | 0.7 | 0.6 |
| 90 | 90 | 20 | 0.5 | 0.5 | 0.5 | 0.5 | 0.5 | 0.5 | 0.5 | 0.5 | 0.5 | 0.5 | 0.5 | 0.5 | 0.5 |
| 90 | 90 | 40 | 0.3 | 0.3 | 0.3 | 0.4 | 0.4 | 0.3 | 0.3 | 0.4 | 0.4 | 0.3 | 0.3 | 0.3 | 0.3 |
| 90 | 135 | 0 | 0.3 | 0.3 | 0.4 | 0.4 | 0.5 | 0.5 | 0.5 | 0.5 | 0.4 | 0.3 | 0.3 | 0.3 | 0.4 |
| 90 | 135 | 20 | 0.3 | 0.3 | 0.3 | 0.3 | 0.4 | 0.4 | 0.4 | 0.4 | 0.3 | 0.3 | 0.3 | 0.3 | 0.3 |
| 90 | 135 | 40 | 0.3 | 0.3 | 0.2 | 0.3 | 0.3 | 0.3 | 0.3 | 0.3 | 0.2 | 0.2 | 0.3 | 0.3 | 0.3 |
| 90 | 180 | 0 | 0.3 | 0.3 | 0.2 | 0.2 | 0.3 | 0.3 | 0.3 | 0.3 | 0.2 | 0.2 | 0.3 | 0.3 | 0.3 |
| 90 | 180 | 20 | 0.3 | 0.3 | 0.2 | 0.2 | 0.2 | 0.2 | 0.2 | 0.2 | 0.2 | 0.2 | 0.3 | 0.3 | 0.2 |
| 90 | 180 | 40 | 0.3 | 0.3 | 0.2 | 0.2 | 0.2 | 0.2 | 0.2 | 0.2 | 0.2 | 0.2 | 0.3 | 0.3 | 0.2 |

## Appendix II

# System Sizing Worksheets

# WORKSHEET #1: DEFINE SITE CONDITIONS AND SOLAR AVAILABILITY

| SYSTEM: |
|---|

| | | |
|---|---|---|
| SYSTEM LOCATION: | LATITUDE: | LONGITUDE: |
| INSOLATION LOCATION: | LATITUDE: | LONGITUDE: |

| MONTH | Location | | | Array plane | | | | | |
|---|---|---|---|---|---|---|---|---|---|
| | Ambient temperature | Horizontal insolation | * | tilt, azimuth, shadow Factor (appendix I) | = | Insolation | | | |
| | °C | $\frac{kWh}{m^2day}$ | * | fraction | = | $\frac{kWh}{m^2day}$ | * | = | $\frac{kWh}{m^2month}$ |
| January | | | * | | = | | *31 | = | |
| February | | | * | | = | | *28 | = | |
| March | | | * | | = | | *31 | = | |
| April | | | * | | = | | *30 | = | |
| May | | | * | | = | | *31 | = | |
| June | | | * | | = | | *30 | = | |
| July | | | * | | = | | *31 | = | |
| August | | | * | | = | | *31 | = | |
| September | | | * | | = | | *30 | = | |
| October | | | * | | = | | *31 | = | |
| November | | | * | | = | | *30 | = | |
| December | | | * | | = | | *31 | = | |

S = Annual insolation on PV array [kWh/m$^2$] = Σ =

# WORKSHEET #2: ESTIMATE LOADS

| Load Description | AC or DC | AC loads (1) [W] | Inverter efficiency (2) [%] | DC load (3)=(1)/(2) [W] | Duty cycle (4) [h/day] | Duty cycle (5) [day/week] | Daily load (6)= (3)*(4)*(5)/7 [Wh/day] | Nominal voltage (7) [V] | Ah-Load (8)=(6)/(7) [Ah/day] |
|---|---|---|---|---|---|---|---|---|---|
| | | | | | | | | | |
| | | | | | | | | | |
| | | | | | | | | | |
| | | | | | | | | | |
| | | | | | | | | | |
| | | | | | | | | | |
| | | | | | | | | | |
| | | | | | | | | | |
| | | | | | | | | | |
| | | | | | | | | | |
| | | | | | | | | | |
| | | | | | | | | | |
| | | | | | | | | | |
| | | | | | | | | | |
| | | MAXIMUM DC LOAD (9) [W] | | | TOTAL DAILY LOADS (10)=Σ(6) [Wh/day] | | | TOTAL LOAD (11)=Σ(8) [Ah/day] | |

| | | |
|---|---|---|
| DESIGN LOAD (Total load=(11)) | | Ah/Day |
| DESIGN PEAK CURRENT DRAW (Maximum DC load / Nominal Voltage) | | A |
| ANNUAL LOAD ENERGY (Total daily loads * 0,365) | | kWh |

# WORKSHEET #3: GRID CONNECTED SYSTEM (part I)

| Chosen PV array power $P_{PV}$ [$kW_P$] | / | PV efficiency (table 17.2) $\eta_{PV}$ [fraction] | = | PV array area $A_{PV}$ [$m^2$] |
|---|---|---|---|---|
| | / | | = | |

| Chosen PV Array power $P_{PV}$ [$kW_p$] | * | Annual insolation on PV array (worksheet #1) S [$kWh/m^2$] | * | BOS efficiency (see below) $\eta_{BOS}$ [fraction] | * | $K_{PV}$ factor [fraction] | = | Annual produced PV energy $E_{PV}$ [kWh] |
|---|---|---|---|---|---|---|---|---|
| | * | | * | | * | 0.9 | = | |

| Annual produced PV energy $E_{PV}$ [kWh] | / | Annual load energy (worksheet #2) [kWh] | = | PV/load ratio [fraction] | from Figure 17.3 | Directly used PV energy [fraction] |
|---|---|---|---|---|---|---|
| | / | | = | | ==> | |

| Chosen PV Array power $P_{PV}$ [$kW_p$] | * | Optimum inverter size (from table 17.1) [fraction] | = | Inverter nominal power [kW] |
|---|---|---|---|---|
| | * | | = | |

---

| average inverter efficiency [fraction] | * | wiring loss factor (1-loss fraction) [fraction] | = | BOS efficiency $\eta_{BOS}$ [fraction] |
|---|---|---|---|---|
| | * | | = | |

# WORKSHEET #3: GRID CONNECTED SYSTEM (part II)

GENERAL INFORMATION

Utility name: ______________________

Contact address: ______________________

Phone number: ______________________

METERING OPTIONS

Single net metering: ______________ (Y/N)

Size restriction: ______________ (kW)

Dual metering: ______________ (Y/N)

Simultaneous buy/sell: ______________ (Y/N)

Buyback ratio: ______________

SPECIAL REQUIREMENTS

Outdoor PV disconnect? ______________ (Y/N)

| Price paid for sold PV energy [US-$/kWh] | * | Sold PV energy fraction (1-directly used) [fraction] | * | Annual produced PV energy [kWh] | = | Annual income from sold PV energy [US-$] |
|---|---|---|---|---|---|---|
| | * | | * | | = | |

# WORKSHEET #4: SIZE BATTERY BANK

| Design load (worksheet #2) [Ah/day] | * | Days of autonomy (see table below) [Days] | / | Max depth of discharge [fraction] | = | Usable battery capacity [Ah] |
|---|---|---|---|---|---|---|
| | * | | / | | = | |

| | |
|---|---|
| OPERATING TEMP = | [degrees C] |
| DISCHARGE RATE = | 24 x DAYS OF AUTONOMY |
| = | [h] |

| Usable battery capacity [Ah] | / | Usable fraction of capacity available [from graph below] | = | Design battery capacity [Ah] |
|---|---|---|---|---|
| | / | | = | |

RECOMMENDED DAYS OF AUTONOMY

| Northern Latitude | Summer months 5,6,7,8 (days) | Spring/ autumn months 3,4,9,10 (days) | Winter months 11,12,1,2 (days) |
|---|---|---|---|
| 30° | 2 - 4 | 3 - 4 | 4 - 6 |
| 40° | 2 - 4 | 4 - 6 | 6 - 10 |
| 50° | 2 - 4 | 6 - 8 | 10 - 15 |
| 60° | 3 - 5 | 8 - 12 | 15 - 25 |
| 70° | 3 - 5 | 10 - 14 | 20 - 35 |

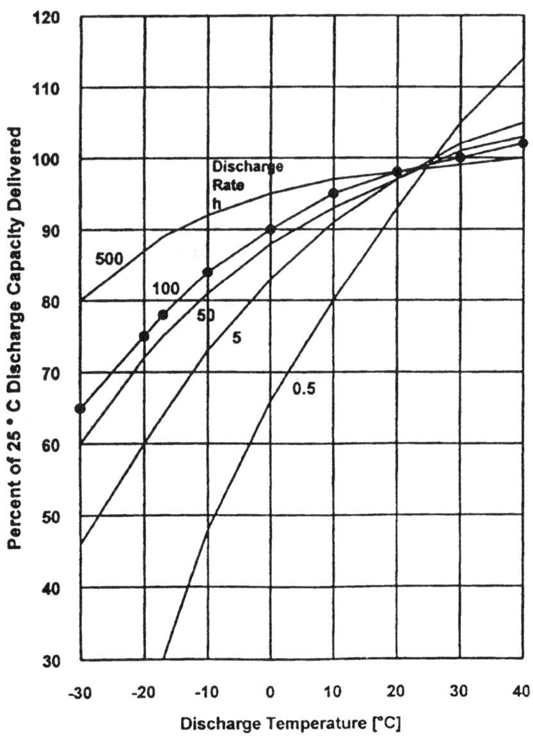

# WORKSHEET #5: SIZE ARRAY & COMPONENTS

| OPERATING SEASON (Months) | |
|---|---|

| Design* month daily load [kWh/day] | / | Lowest** insolation on PV array (worksheet #1) [(kWh/m²)/day] | / | Wiring loss factor (1-loss fraction) [fraction] | / | Charge regulator efficiency [fraction] | / | Battery efficiency [fraction] | = | Design PV array power [$kW_p$] |
|---|---|---|---|---|---|---|---|---|---|---|
| | / | | / | | / | | / | | = | |

| Design PV array power [$W_p$] | * | PV array sizing safety factor (see table below) | = | PV array power [$W_p$] |
|---|---|---|---|---|
| | * | | = | |

| PV array power [$W_p$] | / | Nominal voltage [V] | = | Design array current [A] |
|---|---|---|---|---|
| | / | | = | |

| Design array current [A] | = | Design power conditioner current [A]*** |
|---|---|---|
| | = | |

| PV array power $P_{PV}$ [$kW_p$] | / | PV module efficiency (from table 17.2) $\eta_{PV}$ [fraction] | = | PV array area $A_{PV}$ ($m^2$) |
|---|---|---|---|---|
| | / | | = | |

* When load is constant through the year, the chosen design month is the month with lowest radiation. Otherwise the month is chosen so that the mismatch between the monthly load and insolation on PV surface is the largest (worksheets #1 and #2)

** Insolation on array plane [(kWh/m²)/day] = Peak sun hours [h/day], see definitions

*** If the peak load current is higher than the design array current the power conditioner must be sized on that basis.

PV ARRAY SAFETY FACTOR

| Latitude | Summer [fraction] | Spring/ autumn [fraction] | Winter [fraction] |
|---|---|---|---|
| 30° | 1.1 - 1.3 | 1.1 - 1.4 | 1.2 - 1.6 |
| 40° | 1.1 - 1.4 | 1.2 - 1.4 | 1.3 - 1.7 |
| 50° | 1.2 - 1.5 | 1.3 - 1.6 | 1.4 - 1.8 |
| 60° | 1.2 - 1.6 | 1.3 - 1.7 | 1.4 - 2.0 |

# WORKSHEET #6: CONSIDER HYBRID

| Design array power [$W_p$] | / | Total daily load (worksheet #2) [Wh/day] | = | Array/load ratio [$W_p$/(Wh/day)] |
|---|---|---|---|---|
| | / | | = | |

**HYBRID DESIGN (from graph)**

| YES | NO |
|---|---|
| | |

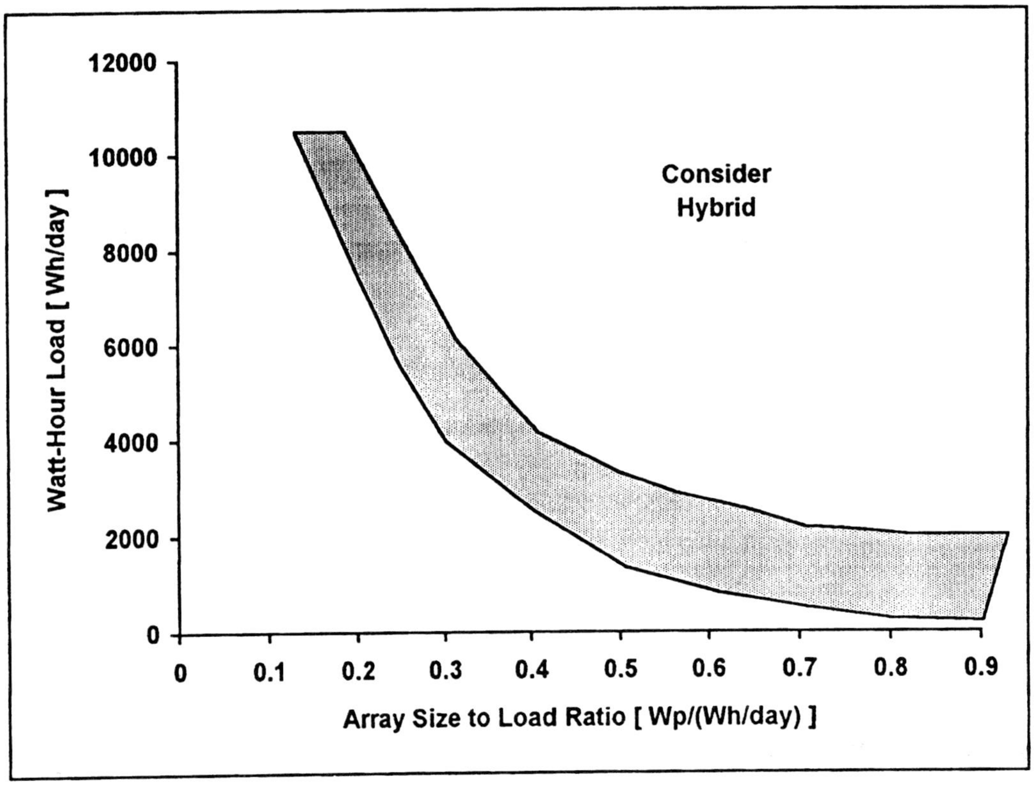

# WORKSHEET #7: SIZE HYBRID

| | | |
|---|---|---|
| PV DESIGN PERIOD | | Months |
| LOWEST INSOLATION DURING PV DESIGN PERIOD (from worksheet #1) | | kWh/m²day |

| Design* month daily load (worksheet #2) [kWh/day] | / | Lowest** insolation on PV array during PV design period (worksheet #1) [(kWh/m²)/day] | / | Wiring loss factor (1-loss fraction) [fraction] | / | Charge regulator efficiency [fraction] | / | Battery efficiency [fraction] | = | PV array power [$kW_p$] |
|---|---|---|---|---|---|---|---|---|---|---|
| | / | | / | | / | | / | | = | |

| PV array power [$W_p$] | / | Nominal voltage [V] | = | Design array current [A] |
|---|---|---|---|---|
| | / | | = | |

| Design array current [A] | * | Lowest** insolation on PV array (worksheet #1) [(kWh/m²)/day] | = | PV load contribution [Ah/day] |
|---|---|---|---|---|
| | * | | = | |

| Design load (worksheet #2) [Ah/day] | * | Days of*** autonomy [days] | / | Maximum depth of discharge [fraction] | / | Usable fraction of battery capacity available (from worksheet #4) | = | Design battery capacity [Ah] |
|---|---|---|---|---|---|---|---|---|
| | * | | / | | / | | = | |

| Design load (worksheet #2) [Ah/day] | / | Battery efficiency [fraction] | / | Rectifier efficiency [fraction] | * | Nominal voltage [V] | = | Design generator load [Wh/day] |
|---|---|---|---|---|---|---|---|---|
| | / | | / | | * | | = | |

| Design generator load [Wh/day] | * | Days of autonomy (see above) [days] | / | Charge time [h] | = | Nominal generator capacity [W] |
|---|---|---|---|---|---|---|
| | * | | / | | = | |

* Load during PV design period, when usually diesel generator is not designed to operate
** Insolation on array plane [(kWh/m²)/day] = Peak sun hours [h/day], see definitions
*** Recommended days of autonomy is 2 - 4 days for all locations.

## Appendix III

# Wire Sizing Tables

The following tables give the maximum distance allowed for selected copper wire sizes and currents. The tables are for 12 V, 48 V, and 120 V systems and the voltage drop is limited to 3%. The wire size is in AWG (American Wire Gage) with its equivalent in wire section in square millimetres ($mm^2$). These tables can be adjusted to reflect a different voltage drop percentage or different wire sections which are not noted in the table by using simple ratios. For example, a 5% table can be calculated by multiplying the values in the table by 5/3, or a 10 $mm^2$ section can be calculated by dividing any column by its section size and multiplying by 10.

The tables are calculated for one-way distance, taking into account that the circuit has two conductors to go from the source of current to the device being powered. For example, if a 48 V array is 10 metres from a 1200 W inverter, the table will show that a No. 12 wire size can be used up to a distance of 10.8 metres.

Note that when sizing the wire, the total current carrying capability (ampacity) of the wire must not be exceeded. Ampacity depends on wire type and temperature. Some ampacity values for copper wire are given below. For other types, please refer to your national electric codes.

| AWG | T, TW, UF | RHW, THW, THHN |
|---|---|---|
| 14 | 15 | 15 |
| 12 | 20 | 20 |
| 10 | 30 | 30 |
| 8 | 40 | 50 |
| 6 | 55 | 65 |
| 4 | 70 | 85 |
| 2 | 95 | 115 |
| 1/0 | 125 | 150 |
| 2/0 | 145 | 175 |
| 3/0 | 165 | 200 |

The following copper wire types are commonly used in PV systems:

**UF:** (sunlight resistant): used for array wiring and underground burial.

**T:** commonly called tray cable and used for array wiring, not for burial.

**TW/THW:** used for interconnecting BOS, must be installed in conduit.

**THHN:** used as battery cables.

**Conductor size for 3 % drop in Voltage**

| **Voltage:** | **12** | **Voltage Drop:** | | **3%** | | | | | | | |
|---|---|---|---|---|---|---|---|---|---|---|---|
| **AWG** | | 14 | 12 | 10 | 8 | 6 | 4 | 2 | 1/0 | 2/0 | 3/0 |
| **Section (mm2)** | | 2.08 | 3.31 | 5.27 | 8.3 | 13.3 | 21.1 | 33.6 | 53.5 | 67.4 | 85.6 |
| **Amps** | **Watts** | Wire Distance from Source of Current to Device (One-Way) - Meters | | | | | | | | | |
| 1 | 12 | 16.9 | 26.9 | 42.8 | 67.5 | 108.1 | 171.5 | 273.1 | 434.9 | 547.8 | 695.8 |
| 2 | 24 | 8.5 | 13.5 | 21.4 | 33.7 | 54.1 | 85.8 | 136.6 | 217.4 | 273.9 | 347.9 |
| 4 | 48 | 4.2 | 6.7 | 10.7 | 16.9 | 27.0 | 42.9 | 68.3 | 108.7 | 137.0 | 173.9 |
| 6 | 72 | 2.8 | 4.5 | 7.1 | 11.2 | 18.0 | 28.6 | 45.5 | 72.5 | 91.3 | 116.0 |
| 8 | 96 | 2.1 | 3.4 | 5.4 | 8.4 | 13.5 | 21.4 | 34.1 | 54.4 | 68.5 | 87.0 |
| 10 | 120 | 1.7 | 2.7 | 4.3 | 6.7 | 10.8 | 17.2 | 27.3 | 43.5 | 54.8 | 69.6 |
| 15 | 180 | 1.1 | 1.8 | 2.9 | 4.5 | 7.2 | 11.4 | 18.2 | 29.0 | 36.5 | 46.4 |
| 20 | 240 | - | 1.3 | 2.1 | 3.4 | 5.4 | 8.6 | 13.7 | 21.7 | 27.4 | 34.8 |
| 25 | 300 | - | - | 1.7 | 2.7 | 4.3 | 6.9 | 10.9 | 17.4 | 21.9 | 27.8 |
| 30 | 360 | - | - | 1.4 | 2.2 | 3.6 | 5.7 | 9.1 | 14.5 | 18.3 | 23.2 |
| 40 | 480 | - | - | - | 1.7 | 2.7 | 4.3 | 6.8 | 10.9 | 13.7 | 17.4 |
| 50 | 600 | - | - | - | 1.3 | 2.2 | 3.4 | 5.5 | 8.7 | 11.0 | 13.9 |
| 100 | 1200 | - | - | - | - | - | - | 2.7 | 4.3 | 5.5 | 7.0 |
| 150 | 1800 | - | - | - | - | - | - | - | 2.9 | 3.7 | 4.6 |
| 200 | 2400 | - | - | - | - | - | - | - | - | 2.7 | 3.5 |

| **Voltage:** | **24** | **Voltage Drop:** | | **3%** | | | | | | | |
|---|---|---|---|---|---|---|---|---|---|---|---|
| **AWG** | | 14 | 12 | 10 | 8 | 6 | 4 | 2 | 1/0 | 2/0 | 3/0 |
| **Section (mm2)** | | 2.08 | 3.31 | 5.27 | 8.3 | 13.3 | 21.1 | 33.6 | 53.5 | 67.4 | 85.6 |
| **Amps** | **Watts** | Wire Distance from Source of Current to Device (One-Way) - Meters | | | | | | | | | |
| 1 | 24 | 33.8 | 53.8 | 85.7 | 134.9 | 216.2 | 343.0 | 546.2 | 869.7 | 1095.7 | 1391.5 |
| 2 | 48 | 16.9 | 26.9 | 42.8 | 67.5 | 108.1 | 171.5 | 273.1 | 434.9 | 547.8 | 695.8 |
| 4 | 96 | 8.5 | 13.5 | 21.4 | 33.7 | 54.1 | 85.8 | 136.6 | 217.4 | 273.9 | 347.9 |
| 6 | 144 | 5.6 | 9.0 | 14.3 | 22.5 | 36.0 | 57.2 | 91.0 | 145.0 | 182.6 | 231.9 |
| 8 | 192 | 4.2 | 6.7 | 10.7 | 16.9 | 27.0 | 42.9 | 68.3 | 108.7 | 137.0 | 173.9 |
| 10 | 240 | 3.4 | 5.4 | 8.6 | 13.5 | 21.6 | 34.3 | 54.6 | 87.0 | 109.6 | 139.2 |
| 15 | 360 | 2.3 | 3.6 | 5.7 | 9.0 | 14.4 | 22.9 | 36.4 | 58.0 | 73.0 | 92.8 |
| 20 | 480 | - | 2.7 | 4.3 | 6.7 | 10.8 | 17.2 | 27.3 | 43.5 | 54.8 | 69.6 |
| 25 | 600 | - | - | 3.4 | 5.4 | 8.6 | 13.7 | 21.8 | 34.8 | 43.8 | 55.7 |
| 30 | 720 | - | - | 2.9 | 4.5 | 7.2 | 11.4 | 18.2 | 29.0 | 36.5 | 46.4 |
| 40 | 960 | - | - | - | 3.4 | 5.4 | 8.6 | 13.7 | 21.7 | 27.4 | 34.8 |
| 50 | 1200 | - | - | - | 2.7 | 4.3 | 6.9 | 10.9 | 17.4 | 21.9 | 27.8 |
| 100 | 2400 | - | - | - | - | - | - | 5.5 | 8.7 | 11.0 | 13.9 |
| 150 | 3600 | - | - | - | - | - | - | - | 5.8 | 7.3 | 9.3 |
| 200 | 4800 | - | - | - | - | - | - | - | - | 5.5 | 7.0 |

**Conductor size for 3 % drop in Voltage**

| Voltage: | 48 | Voltage Drop: | 3% | | | | | | | |
|---|---|---|---|---|---|---|---|---|---|---|
| AWG | | 14 | 12 | 10 | 8 | 6 | 4 | 2 | 1/0 | 2/0 | 3/0 |
| Section (mm2) | | 2.08 | 3.31 | 5.27 | 8.3 | 13.3 | 21.1 | 33.6 | 53.5 | 67.4 | 85.6 |
| Amps | Watts | Wire Distance from Source of Current to Device (One-Way) - Meters | | | | | | | | | |
| 1 | 120 | 67.6 | 107.6 | 171.3 | 269.9 | 432.4 | 686.0 | 1092.4 | 1739.4 | 2191.3 | 2783.1 |
| 2 | 240 | 33.8 | 53.8 | 85.7 | 134.9 | 216.2 | 343.0 | 546.2 | 869.7 | 1095.7 | 1391.5 |
| 4 | 480 | 16.9 | 26.9 | 42.8 | 67.5 | 108.1 | 171.5 | 273.1 | 434.9 | 547.8 | 695.8 |
| 6 | 720 | 11.3 | 17.9 | 28.6 | 45.0 | 72.1 | 114.3 | 182.1 | 289.9 | 365.2 | 463.8 |
| 8 | 960 | 8.5 | 13.5 | 21.4 | 33.7 | 54.1 | 85.8 | 136.6 | 217.4 | 273.9 | 347.9 |
| 10 | 1200 | 6.8 | 10.8 | 17.1 | 27.0 | 43.2 | 68.6 | 109.2 | 173.9 | 219.1 | 278.3 |
| 15 | 1800 | 4.5 | 7.2 | 11.4 | 18.0 | 28.8 | 45.7 | 72.8 | 116.0 | 146.1 | 185.5 |
| 20 | 2400 | - | 5.4 | 8.6 | 13.5 | 21.6 | 34.3 | 54.6 | 87.0 | 109.6 | 139.2 |
| 25 | 3000 | - | - | 6.9 | 10.8 | 17.3 | 27.4 | 43.7 | 69.6 | 87.7 | 111.3 |
| 30 | 3600 | - | - | 5.7 | 9.0 | 14.4 | 22.9 | 36.4 | 58.0 | 73.0 | 92.8 |
| 40 | 4800 | - | - | - | 6.7 | 10.8 | 17.2 | 27.3 | 43.5 | 54.8 | 69.6 |
| 50 | 6000 | - | - | - | 5.4 | 8.6 | 13.7 | 21.8 | 34.8 | 43.8 | 55.7 |
| 100 | 12000 | - | - | - | - | - | - | 10.9 | 17.4 | 21.9 | 27.8 |
| 150 | 18000 | - | - | - | - | - | - | - | 11.6 | 14.6 | 18.6 |
| 200 | 24000 | - | - | - | - | - | - | - | - | 11.0 | 13.9 |

| Voltage: | 120 | Voltage Drop: | 3% | | | | | | | | |
|---|---|---|---|---|---|---|---|---|---|---|---|
| AWG | | 14 | 12 | 10 | 8 | 6 | 4 | 2 | 1/0 | 2/0 | 3/0 |
| Section (mm2) | | 2.08 | 3.31 | 5.27 | 8.3 | 13.3 | 21.1 | 33.6 | 53.5 | 67.4 | 85.6 |
| Amps | Watts | Wire Distance from Source of Current to Device (One-Way) - Meters | | | | | | | | | |
| 1 | 120 | 169.1 | 269.0 | 428.4 | 674.6 | 1081.0 | 1715.0 | 2731.0 | 4348.5 | 5478.3 | 6957.7 |
| 2 | 240 | 84.5 | 134.5 | 214.2 | 337.3 | 540.5 | 857.5 | 1365.5 | 2174.3 | 2739.2 | 3478.8 |
| 4 | 480 | 42.3 | 67.3 | 107.1 | 168.7 | 270.3 | 428.8 | 682.8 | 1087.1 | 1369.6 | 1739.4 |
| 6 | 720 | 28.2 | 44.8 | 71.4 | 112.4 | 180.2 | 285.8 | 455.2 | 724.8 | 913.1 | 1159.6 |
| 8 | 960 | 21.1 | 33.6 | 53.5 | 84.3 | 135.1 | 214.4 | 341.4 | 543.6 | 684.8 | 869.7 |
| 10 | 1200 | 16.9 | 26.9 | 42.8 | 67.5 | 108.1 | 171.5 | 273.1 | 434.9 | 547.8 | 695.8 |
| 15 | 1800 | 11.3 | 17.9 | 28.6 | 45.0 | 72.1 | 114.3 | 182.1 | 289.9 | 365.2 | 463.8 |
| 20 | 2400 | - | 13.5 | 21.4 | 33.7 | 54.1 | 85.8 | 136.6 | 217.4 | 273.9 | 347.9 |
| 25 | 3000 | - | - | 17.1 | 27.0 | 43.2 | 68.6 | 109.2 | 173.9 | 219.1 | 278.3 |
| 30 | 3600 | - | - | 14.3 | 22.5 | 36.0 | 57.2 | 91.0 | 145.0 | 182.6 | 231.9 |
| 40 | 4800 | - | - | - | 16.9 | 27.0 | 42.9 | 68.3 | 108.7 | 137.0 | 173.9 |
| 50 | 6000 | - | - | - | 13.5 | 21.6 | 34.3 | 54.6 | 87.0 | 109.6 | 139.2 |
| 100 | 12000 | - | - | - | - | - | - | 27.3 | 43.5 | 54.8 | 69.6 |
| 150 | 18000 | - | - | - | - | - | - | - | 29.0 | 36.5 | 46.4 |
| 200 | 24000 | - | - | - | - | - | - | - | - | 27.4 | 34.8 |

**Note: 1 meter=** 3.281 **ft**

## Appendix IV

# Tender Documents

# Tender Document for Solar Modules

## 1. General

### 1.1. Product

product's name ________ type: ________

manufacturer/supplier ________

adress ________

telephone/fax ________

### 1.2. Warranty

general warranty ________ years power warranty ________ years

peak power at purchase ________ Wp and after warranty time ________ Wp

### 1.3. Power approval (final inspection)

The final inspection is done by measuring power according to the ISPRA-guidelines.

### 1.4. Documentation

The product offered should be sufficiently documented.
The following are considered to be required:

data sheet
current-voltage-characteristics curve under STC and NOCT
dimensions and weights
information on connection box and laminate composition
efficiency as a function of irradiation, temperature and irradiation angle

## 2. Specifications

### 2.1. Electrical data

2.1.1 peak power ________ Wp nominal power under STC: ________ Wp

2.1.2 voltage

max. operation voltage ________ VDC max. open circuit voltage ________ VDC

voltage under STC ________ VDC MPP-voltage ________ VDC

2.1.3 current

MPP-current ________ A short circuit current ________ A

2.1.4 leak current ________ $\mu$A

2.1.5 module variation

power max. value ________ Wp min. value ________ Wp

voltage max. value ________ VDC min. value ________ VDC

current max. value ________ A min. value ________ A

2.1.6 efficiency under STC ________ %

2.1.7 temperature dependence

voltage coefficient ________ %/K current coefficient ________ %/K

power coefficient ________ %/K

2.1.5 protecting diodes

bypass diodes included ________ yes/no number: ________

type of bypass diodes ________

2.1.6 connections Description/plan of the electrical connections

2.1.7 classification
are the delivered
modules classified? ______ yes/no number of cl.: ______

2.1.8 material of the cell
monocrystalline, polycrystalline, amorphous or others? ______

**2.2 Mechanical data**

2.2.1 dimensions
width, length, depth , weight mm ______ mm ______ mm ______ kg

2.2.2 material choice
colour ______
frame construction ______

2.2.3 attachement points The modules must have suitable points for attachement

2.2.4 grade of reflection ______

2.2.5 module tests
passed module tests, z. B. ESTI, JPL-Block V

______

**2.3. Physical data**

2.3.1 vapour diffusion ______ mg/(m*hPa)

2.3.2 heat conductivity ______ W/(m*K)

2.3.3. mechanical strength
admitted forces ______ ______ N
______ ______ N
______ ______ N
hail resistance ______
fire reference no. ______

2.3.4 Dilatation
dilatation coefficient ______ /K

2.3.5 corrosion proof
salt water ______
solvents ______
other substances ______

2.3.6. operation temperature
temperature under STC ______ °C

## 3. Costs

cost per modul ______ valuta
cost per power unit ______ valuta/Wp
transport included ______ yes/no
taxes ______
payment conditions ______

**Place, date:** ______

**Signature:** ______

# Tender Document for DC-AC-Inverter

## 1. General

### 1.1. Product

product's name ________ technology: ________

manufacturer/supplier ________

adress ________

telephone/fax ________

### 1.2. Warranty

warranty time ______ years power warranty ______ years

### 1.3. Power approval (final inspection)

The final inspection is done by testing efficiency, MPPT-operation and control of several operation stages.

### 1.4. Documentation

The product offered should be sufficiently documented.
The following are considered to be required:

data sheets
curves of efficiency versus DC-input voltage and power level
list of fulfilled norms and regulations

## 2. Specifications

### 2.1. DC-input

2.1.1 power

DC-nominal power ______ W
max. allowed DC-power ______ W min. DC-power for Startup ______ W

2.1.2 voltage

nominal voltage ______ VDC MPT-range ______ VDC-VDC
isolation test voltage ______ VDC open circuit voltage ______ VDC

2.1.3 normal current ______ A

2.1.4 current ripple ______ % peak-peak

2.1.5 terminal ______ mm2-mm2

2.1.6 interferrence suppression under ________

### 2.2. AC-output

2.2.1 power

nominal power ______ W

2.2.2 voltage

nominal voltage ______ VAC output voltage range ______ VAC-VAC
isolation test voltage ______ VAC

2.2.3 current harmonics ______ %total possible grid impedance ______

mains frequency ______ Hz mains frequency range ______ (+/-) Hz

2.2.4 cos phi under peak power ______

2.2.5 terminal ______ mm2-mm2

2.2.6 overvoltage protection level ________

**2.3 Efficiency**

at 10% of nominal DC power: __________ %

at 50% of nominal DC power: __________ %

at 100% of nominal DC power: __________ %

no load losses __________ W

stand-by losses __________ W

**2.4 Mechanical data**

2.4.1 dimensions

width, length, depth mm ________ mm ________ mm

weight ________ kg

2.4.2 cooling

type of cooling ____________________

special cooling necessary? ____________________

2.4.3 attachement points inverter must have suitable points for attachement

________________________________________

**2.5. System data**

2.5.1 environmental data

temperature range min.________ °C max.________ °C
humidity range min.________ % max.________ %
noise production min.________ dB max.________ dB

2.5.2 control

operating at ________ VDC shutting down at ________ VDC
reactions to DC-power overload ____________________
reactions to AC-voltage break ____________________
starting up automatically ________ yes/no time to start ________
connectable to PC ________ yes/no type of connection ________
software ____________________ operating system ________

2.5.3 operation of several inverters together / master - slave
possibility ________ yes/no control ________

2.5.4 safety measures
overcurrent protection devices ____________________
possibilities for emergency shut down ____________________ yes/no
control of leak current in the pv-field ____________________ yes/no

## 3. Costs

costs ________ valuta
costs per power unit ________ valuta/Wp
transport included ________ yes/no
taxes ____________________
payment conditions ____________________

**Place, date:** ____________________

**Signature:** ____________________

## Appendix V

# Maintenance Logsheets

In this appendix a collection of logsheets is given to support the systematic maintenance of PV modules, batteries, power conditioning system etc.

It is very important to have each one of the PV modules as well as the elements of the battery bank well identified. Also, the different blocks or cards in the power conditioning or regulator system should be marked clearly.

For the power conditioning subsystem and/or regulator, the most important thing to observe is the good performance of the equipment in reference with the rest of the subsystems: PV modules, battery and loads. For this it is very useful to have volt meters and ampere meters in the DC side as well as in the AC side. To help the specialist in case of failure or bad performance, it is convenient to have the connection schemes of the equipment on hand, in which every electronic card should be well identified. Of course for all cases it is essential to have the schematic of electrical connections of the whole PV system.

**PV-SYSTEM MAINTENANCE**
**SHEET No 1**

**PV Array**

| | OK ? | |
|---|---|---|
| DATE OF CHECK: | | |
| NAME: | yes | no |
| **1.1 FRONTAL FACE** | | |
| * Damage/deterioration or decoloration of the encapsulating material | | |
| * Broken cells | | |
| * Cells decoloured | | |
| * Bubbles | | |
| * Humidity | | |
| **1.2 BACK FACE (If it can be surveyed)** | | |
| * Damage of connection cabling | | |
| * Bubbles | | |
| * Damage of connection boxes | | |
| (Choose at random 10% of the boxes installed) | | |
| - Corrosion | | |
| - Humidity | | |
| - Tightening/Screws | | |
| **1.3 CLEANING OF PV MODULES & OTHERS** | | |
| COMMENTS: | | |

**PV-SYSTEM MAINTENANCE**
**SHEET No 2a**

**Battery**

DATE: NAME:

| ELE-MENT NUM-BER | DATE | MONTHLY TESTS | | | |
|---|---|---|---|---|---|
| | | CORROSION OF TERMINALS AND CONNECTORS [YES/NO] | ELECTRO-LYTE LEVEL | ELECTRO-LYTE LEAKAGE [YES/NO] | CELL VOLTAGES [V] |
| | | | | | |
| | | | | | |
| | | | | | |
| | | | | | |
| | | | | | |
| | | | | | |
| | | | | | |
| | | | | | |
| | | | | | |
| | | | | | |
| | | | | | |
| | | | | | |
| | | | | | |
| | | | | | |
| | | | | | |
| | | | | | |
| | | | | | |

**PV-SYSTEM MAINTENANCE**
**SHEET No 2b**

**Battery**

DATE: NAME:

| **ELE-MENT NUM-BER** | **DATE** | **QUARTERLY TESTS** | | | | | |
|---|---|---|---|---|---|---|---|
| | | **CELL VOLT-AGES** **[V]** | **SPECI-FIC GRA-VITY** **[g/cm³]** | **CELL TEMPE-RATU-RES** **[°C]** | **EQUAL-IZING CHARGE** **[YES/NO]** | **CLEAN-ING AND RETIGHT-ENING** **[YES/NO]** | **CELL CRACKS** **[YES/NO]** |
| | | | | | | | |
| | | | | | | | |
| | | | | | | | |
| | | | | | | | |
| | | | | | | | |
| | | | | | | | |
| | | | | | | | |
| | | | | | | | |
| | | | | | | | |
| | | | | | | | |
| | | | | | | | |
| | | | | | | | |
| | | | | | | | |
| | | | | | | | |
| | | | | | | | |
| | | | | | | | |

**PV-SYSTEM MAINTENANCE**
**SHEET No 3**

**Other System Components**

| DATE OF CHECK: | OK ? | |
|---|---|---|
| NAME: | yes | no |
| **3.1 POWER CONDITIONING** | | |
| * Inverter: | | |
| - wiring | | |
| - set points | | |
| * Regulators: | | |
| - wiring | | |
| - set points | | |
| * Controllers: | | |
| - wiring | | |
| - set points | | |
| **3.2 BACKUP GENERATOR** (see manual) | | |
| * Leakages | | |
| * Oil check | | |
| * Start-up | | |
| **3.3. GROUNDING MEASUREMENT [OHMS]** | | |
| **3.4 LIGHTNING PROTECTION** | | |
| COMMENTS: | | |

## Appendix VI

# Trade-Off Considerations

*Figure VI.1 The Norwegian Low Energy Dwelling. A three apartment row house with different solutions of the use of solar energy. PV roof in the middle, solar thermal roof panel to the right and other low energy properties (not shown on picture) to the left.*

Solar energy is always used in a building in one form or another. i.e. heat gain and daylighting. Extended use of solar energy might also involve photovoltaics and/or solar thermal collectors. Integration of several forms of utilization of solar energy in a building involves considerations of costs, technical performance, legal regulations, safety, environmental aspects and, of course, the architectural aspects. All these considerations must be seen in relation to the needs and wishes of the customer and the designer.

Faced with different proposals for integrating the various uses of solar energy, the decision of choosing between alternatives that must be made by the customer can be a complex task. Decision criteria with respect to the above mentioned aspects are not comparable. Further, the individual priorities may vary strongly from one customer to another, and contradictory preferences may easily occur.

Often, the decision is made in an intuitive way. This may be adequate if the customer either has a good insight into all aspects or he/she has strong preferences leading to obvious choices. In general, however, the choice is not trivial, and a decision-making tool is needed by which a multi-criterion optimization process is made possible. The decision process is initiated by some conflict, usually caused by the inevitable choice between various, completely different alternatives. The goal is to reduce this conflict by finding the alternative that corresponds in the 'best' way to the knowledge and preferences of the decision maker.

The process of decision-making has several stages: conflict, predecision (rules and criteria), partial decisions (alternatives to be analysed), final decision (comparison and selection), post-decision (regret or confidence) and action. All these stages involve elements of uncertainty.

A number of methods can be used in this decision process. Some of these are: Multi Attribute Utility Theory (MAUT), Social Judgement Theory, Compromise Programming, Analytic Hierarchy Process (AHP).

The AHP method, developed by the mathematician Thomas L. Saaty in 1971-75, is very simple and has become popular in multi- criterion decision making. Indeed, this method can be used to decide between different alternatives in which the various uses of solar energy are integrated into buildings.

The method implies only pairwise comparisons between aspects and attributes of the projects that are subject to alternative decisions. The process is broken down into levels, where the top level is the ultimate goal: 'Choose the best project'. Second level is the main aspects to consider, i.e. energy savings, economy, technical issues, architectural features, and environmental considerations. Third level might be the projects themselves between which to choose. Intermediate levels with sub-attributes can be added. Pairwise comparison is then applied between attributes of one level with respect to the aspects of the above level. This manner of quantification can be understood intuitively.

A simple example of use of the AHP method is included here. A software package Expert Choice (version 8) was used. Figure VI.1 shows a three apartment building, the Norwegian Low Energy Dwelling, cf. section 14.14. Rather than the actual solution shown in Figure VI.1, one might consider these three sections of the house as three different projects in which the use of solar energy is different (third level):

Alt. 1: PV
The whole south-facing roof is covered with photovoltaic cell modules

Alt. 2: THERMAL
The whole roof is covered with thermal collectors

Alt. 3: 50/50
PV and thermal collectors share the surface of the roof.

These three alternatives are specified this way only to visualize the AHP method in a simple way. Note that the alternatives do not correspond to the solar energy installations implemented in the Norwegian Low Energy Dwelling as seen in the figure.

The three example projects defined above (third level) are compared using four different criteria (second level): economy, technical performance, environmental friendliness and architectural quality.

The pairwise comparison is given as a number 1 to 9 on a fundamental scale which reflects the relative strength of preference and/or feeling, or a real number reflecting the result of some calculation. The number 1 means the two aspects are of equal importance and the number 9 means the first aspect is of extreme importance over the second. Numbers are reciprocals when the comparison is taken the opposite way.

One example: Thermal collectors are strongly favoured over photovoltaic cells with respect to costs, i.e. the number 7 might be assigned to this pair between level 3 and 2. Further, one decision maker might put a rather strong weight on economy compared to architectural beauty, so the number 5 might be relevant to this pair between level 2 and 1.

Without going into detail, the numbers specified by the user in this manner are input to the Expert Choice software. This software then collects the preference numbers in input matrices. Eigenvectors are calculated, normalized and put into new matrices, which are multiplied giving a vector containing the relative (normalized) priority of the projects at the bottom level.

Having done this, the software offers simple procedures for doing what-if analyses, or sensitivity calculations.

The result can be displayed graphically as shown in Figure VI.2 below.

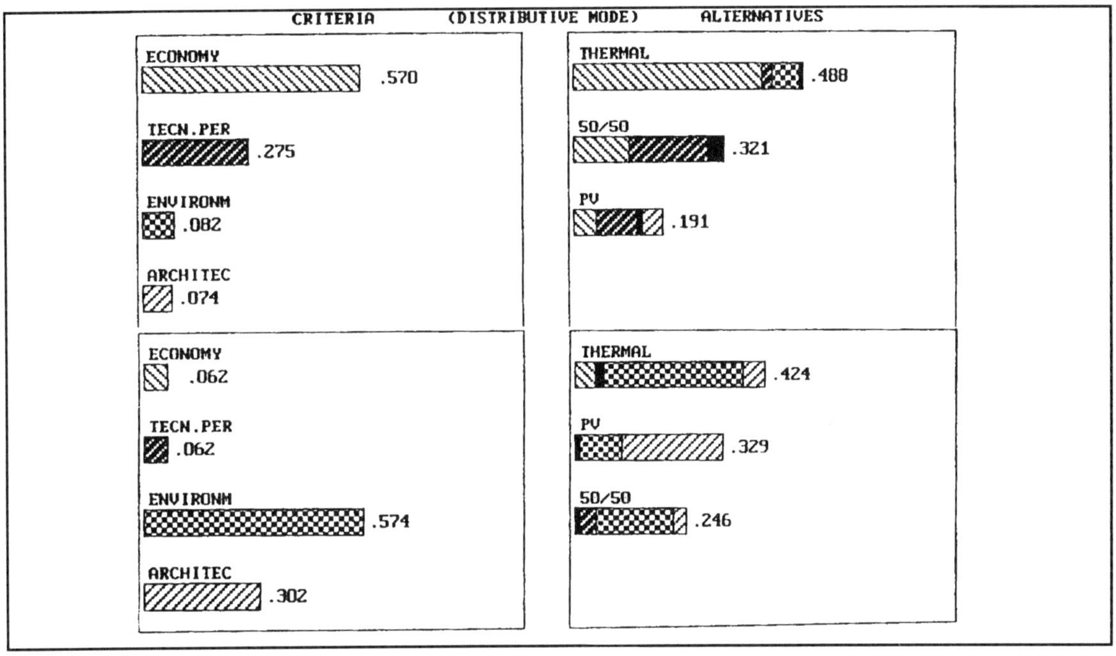


*Figure VI.2* *Graphical representation of criteria and the decision between alternatives for two different types of decision makers. The upper (left) part shows a decision maker that cares about economy and technical performance much more than environmental and architectural aspects. The right part shows the final decision: Solar thermal is favoured. The lower (left) part shows a somewhat contrary opinion. Environmental and architectural aspects are weighted over economy and performance. This changes the ranking between PV and the 50/50 solution, but it still favours the solar thermal alternative.*

The pattern in each of the bars shows the contribution to the final decision from the attributes to the left. This can easily be used interactively to do sensitivity studies (what-if-studies). By changing the weight on the aspects at level 2, the software will immediately show the response to the final decision.

Both these cases ended up with the solar thermal alternative as the favoured choice. Let us now see in what way the weight of the criteria have to be altered in order to end up with the other two alternatives. This effect is shown in Figure VI.3 below.

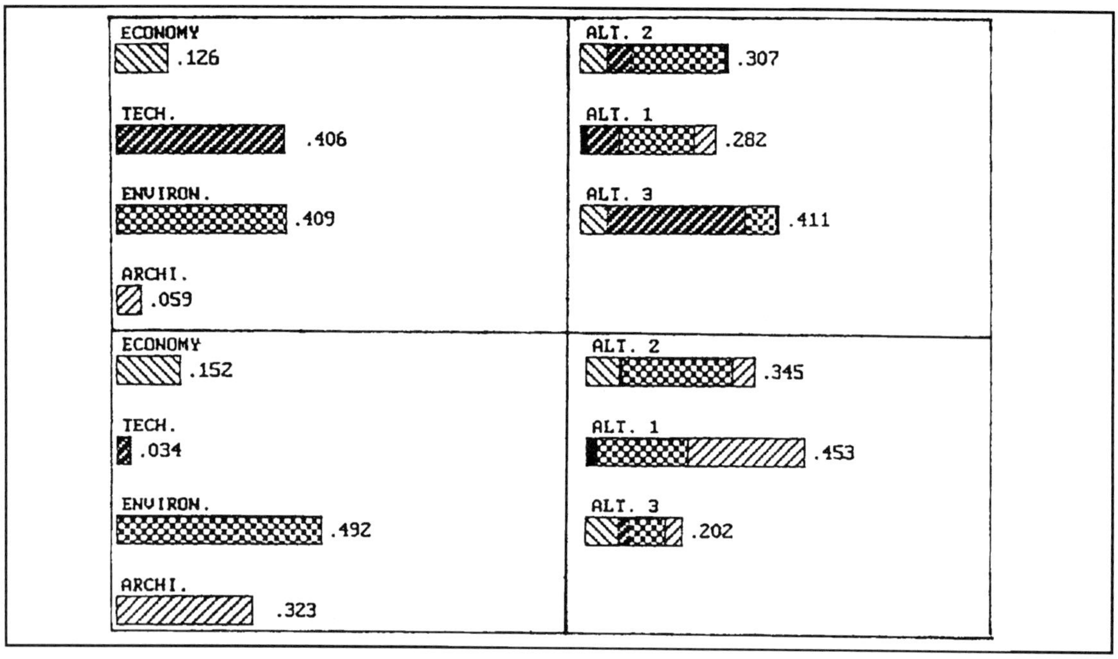


*Figure VI.3* *Sensitivity analysis showing the necessary changes in the weighting profile of the four criteria at level 2. If technical performance is emphasized along with environmental issues at the expense of the other two (economy and architectural features), it is seen from the upper part of the figure that the favoured choice is the mixed system. However, if the importance of technical performance is strongly reduced, and weight is mainly put on environmental and architectural features, it is seen from the lower part of the figure that the PV alternative is favoured over the other two.*

Of course, these results do not hold in general. They are strongly dependent on the underlying technical and economical analysis of the given projects, which results in the preference numbers explained above. This underlying analysis is not shown here. It is, however, emphasized that detailed studies of costs and technical performance using separate simulation software should give realistic and reliable input to the decision tool. Even the environmental and architectural features of a given set of project alternatives could be given a well defined, although sometimes subjective and qualitative value for the alternatives in question.

These rather simple examples show that the AHP method can lead to virtually any conclusion. This might be held up as a disadvantage of the method. In fact, it should be looked upon as an advantage. The reason is that the method offers a simple way of tracing and documenting the user's preferences.

If the conclusion is unexpected, it can be seen from the different steps in the reasoning why that particular alternative was ranked on top, and also what kind of changes in preferences are needed to alter the conclusion. If these alterations seem unreasonable in some way to the user, then even quite unexpected, and even unpopular conclusions must be accepted by the decision maker.

## Appendix VII

# Cost of PV

The economics of grid-connected or stand-alone PV systems is determined by the cost and the revenues for the electricity produced. The cost of a PV system can be grouped into

- capital cost (investments, interest rate, depreciation, replacement),
- consumables (fuel or energy cost),
- operating cost (maintenance, personnel),
- other cost (insurance, taxes).

Often, the last three items are combined and called operating cost.

Based on this different cost, the energy price in US$/kWh can be calculated for a given location (insolation) and a given lifetime. Normally, a lifetime of 25 years is assumed.

The investment cost is mainly made up of the module cost, the cost for the balance of system (BOS) as well as the labour cost for planning and installation. For residential grid-connected systems a typical distribution of cost is shown in Figure VII.1, whereas Figure VII.2 is valid for stand-alone systems with back-up generator.

The absolute value of the initial cost varies over a wide range depending on the individual boundary conditions and the quality of the realized system. For example, in the German '1000 Roofs Programme', the initial cost ranged from US$ 10 to 25 per Wp. For high quality hybrid systems, e.g. energy supplies for alpine mountain huts, the cost is in the range of US$ 20 to 25 per Wp.

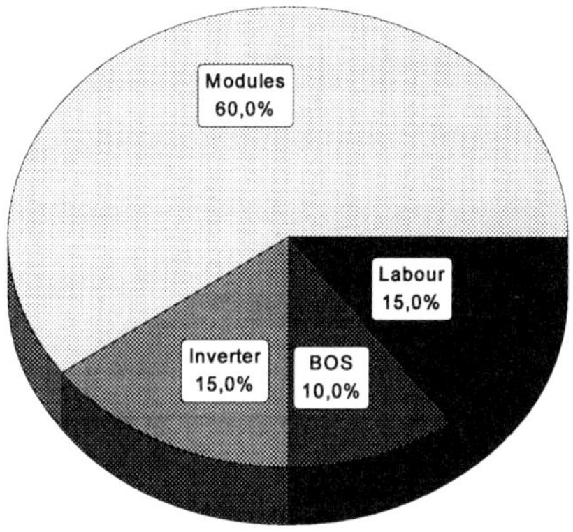


*Figure VII.1* *Distribution of cost for a typical 2 $kW_p$ residential grid-connected PV system.*

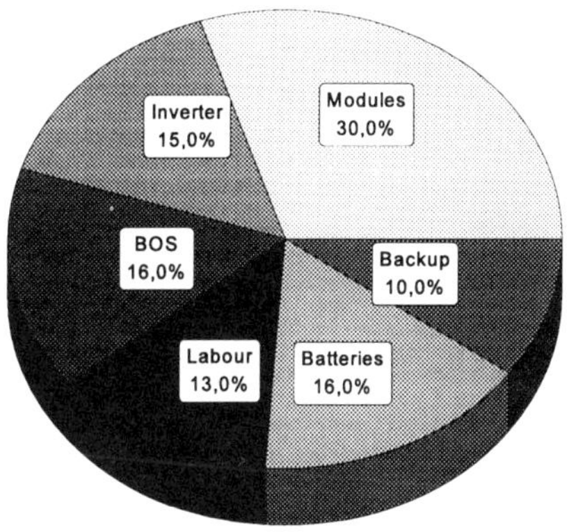


*Figure VII.2* *Distribution of cost for typical 2 $kW_p$ residential hybrid PV system.*

Looking at facades, the cost for a PV facade must be compared with those of a conventional one. Replacing a glass facade by a PV facade leads to additional cost in the range of US$ 1000 to 1500 per $m^2$ including the BOS cost.

In general, it can be expected that the initial cost for PV systems will further decrease in future due to the development of new technologies, improvement of production processes and by increasing the production quantities (see Figure VII.3). Investigations show that the production cost for PV modules in the past has decreased by 15 % whenever the cumulated module production has doubled. Based on this experience and today's production growth rate of 20 % it can be expected that in the year 2010 the average module price will be approximately US$ 2 per Wp[1].

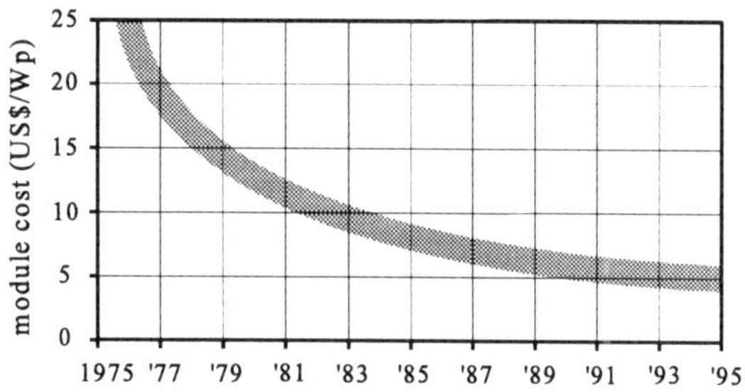


*Figure VII.3 Development of PV power module cost.*

For PV inverters, a price reduction by a factor of 3 can be expected when comparing today's PV inverter cost with that of industrial electronic drives which have similar structures and are produced in large quantities. The same statement applies for charge regulators in the power range of several kilowatts.

Batteries are already mass produced today so that no significant price reduction can be expected. The cost for high quality tubular plate batteries is in the range of US$ 300 per kWh. On the other hand, the life-cycle cost of an autonomous system can be drastically reduced by extending the battery´s lifetime by an improved operating strategy and by appropriate battery peripherals. Experience shows that the battery´s 15 to 20 % share of the initial cost can add up to more than 50 % over the system´s lifetime when taking into account a replacement of the battery every three to five years. Mass production and standardization can further reduce the BOS cost as well as the planning cost.

For residential grid-connected systems the cost of maintenance and replacements can be expected to be approximately 0.5 to 1 % per year of the initial cost. Due to the above mentioned maintenance and replacement of batteries, in stand-alone systems the operation cost depends heavily on the battery's lifetime; 3 to 4 % per year can be anticipated.

When PV systems are dismantled, the modules can be disposed of or recyled like conventional building materials without extra cost. This is true for today's silicon cell technology; however, it will have to be re-examined when new technologies incorporating toxic materials come on the market.

Taking all the above mentioned cost factors into account, for central European conditions the PV energy cost today is in the range of US$ 0.6 to 1.3 per kWh for grid-connected systems and US$ 2.5 to 5 per kWh for stand-alone hybrid systems. Although in the latter case the energy price is much higher than usual, hybrid systems will be cost-competitive with grid extensions or pure diesel-generated supplies. Assuming realistic price reductions for PV modules and BOS, these costs will decrease to US$ 0.3 to 0.6 per kWh, US$ 1.5 to 2.5 per kWh respectifely for stand-alone systems in the year 2005.

It must be pointed out that these cost figures strongly depend on the interest rate due to the high capital cost and the long lifetime of the systems. Furthermore, political decisions and general economic developments have a large influence on this cost.

---

[1]Source: W. Hoffmann, '*Zukunftstechnologie Photovoltaik*', etz, 15/1995.

## Appendix VIII

# Glossary

**Air mass (AM)**
The path length of light through the atmosphere is described in terms of an equivalent relative air mass. AM0 corresponds to the solar spectrum in outer space; at the equator the average spectrum is AM1 and the reference spectrum for STC was defined to be AM1.5 (average spectrum at 45° lattitude).

**Alternating current (AC)**
Electric current in which the direction of flow is reversed at frequent intervals, usually 100 or 120 times per second (50 or 60 cycles per second or 50/60 Hz).

**Amorphous silicon**
Silicon in which the atoms are not arranged in an ordered pattern as in crystalline silicon.

**Ampere (A)**
The unit for the electric current.

**Ampere-hour (Ah)**
Quantity of electricity or measure of charge.
1 Ah = 3600 C (Coulomb).

**Array**
See 'Photovoltaic array'.

**Atrium**
Large, top-lit space rising through several floors in modern buildings (old Roman concept).

**Autonomous system**
A stand-alone photovoltaic system which has no back-up generating source. May or may not include storage batteries.

**Awnings**
Covering, to screen persons or parts of buildings from the sun or rain.

**Back-up generator**
Supplementary electricity source to ensure full-time cost-effective electricity supply.

**Balance of system (BOS)**
The parts of a photovoltaic system other than the array: switches, controls, meters, power conditioning equipment, supporting structure for the array and

storage components, if any.

**Batten** — Small rectangular piece of timber used to provide fixings for tiles or slates. Cover-slip concealing the joint between two boards, or a strip of timber fixed across two parallel boards to join them together.

**Batten seam** — Joint in a roof formed over a wooden strip or roll.

**Battery capacity** — Amount of ampere-hours that can be discharged from the battery under specified conditions of discharge (cut-off voltage, current and temperature).

**Battery charge regulator** — An electrical device used to keep current from running backwards through an array at night or during periods of low sunlight, thereby preventing drainage of the storage battery.

**Blocking diode** — Diode connected in series to a PV string; it protects its modules from a reverse power flow and thus against the risk of thermal destruction of solar cells.

**Breast wall** — Retaining wall, or parapet which is breast-high from the floor.

**Building envelope** — The outside of a building that contains the interior space, including the roof: the skin or waterproof covering of the structure.

**Bypass diode** — Diode connected anti-parallel across a part of the solar cells of a PV module. It protects these solar cells from thermal destruction in case of total or partial shading of individual solar cells whilst other cells are exposed to full light.

**Cell** — See 'Photovoltaic cell'.

**Cladding** — External face or skin of a building.

**Clerestory** — Any window, row of windows, or openings in the upper part of a building.

**Curtain wall** — Non-load-bearing wall placed as a weather-proof membrane round a structure, and usually made of glass or metal.

**Cycle life** — Amount of discharge-charge cycles that a battery can tolerate under specified conditions before it fails to meet specified criteria as to performance (e.g. capacity decreases to 80% of the nominal capacity).

**Direct current (DC)** — Electric current in which electrons flow in one direction only. Opposite of alternating current.

**DC to DC Converter** — Electronic circuit to convert DC-voltages (e.g. PV-module voltage) into other levels (e.g. load voltage). Can be part of MPPT.

**Discharge rate** — The rate, usually expressed in amperes or time, at which electrical current is taken from the battery.

**DOD** — 'Depth-of-Discharge', 100% - SOC (see SOC).

**Electrical grid** — An integrated system of electricity distribution, usually covering a large area.

**Electrolyte** — A liquid conductor of electricity.

**Energy density** — The ratio of the energy available from a battery to its volume (Wh/l) or mass (Wh/kg).

**Flashing** — Piece of metal let into the joints of brickwork to lap over a gutter, or set along the slates of a roof, to prevent water from penetrating at the junctions. To flash is to make water-tight joints.

**Float charge** — Float charge is the voltage required to counteract the self-discharge of the battery at a certain temperature.

**Float life** — Number of years that a battery can keep its stated capacity when it is kept at float charge (see Float charge).

**Gassing current** — Portion of charge current that goes into electrolytical production of hydrogen and oxygen from the electrolytic liquid. This current increases with increasing voltage and temperature.

**Gel-type battery** — Lead-acid battery in which the electrolyte is composed of a silica gel matrix.

**Grid** — See 'Electrical grid'.

**Grid-connected (PV System)**

A PV system in which PV arrays act like central 'generating plants' supplying power to the grid. Either the PV system is operated by the utility, or (in what is known as a grid-interactive system) individual buildings in the grid are equipped with PV systems that feed into the grid when they generate excess power, and draw from the grid at night and in periods of low sunshine.

**Grid-interactive (PV system)**

See 'Grid-connected (PV system)'.

**Hybrid PV system**

A PV system that includes other sources of electricity generation, such as diesel or wind generator.

**Inverter**

A PV inverter is a power converter which transforms DC voltage and current of the PV generator into single or multiphase AC voltage and current.

**IP**

Ingress protection, describes with two figures the protection level against mechanical impact and water penetration.

**ISPRA-Guidelines**

Guidelines for the Assessment of Photovoltaic Plants, Published by the Joint Research Centre of the Commission of the European Communities, Ispra, Italy.

**Junction box**

A PV generator junction box is an enclosure where all PV strings are electrically connected and where protection devices can be located, if necessary.

**Kilowatt-hour (kWh)**

One thousand watts acting over a period of one hour. The kWh is a unit of energy. 1 kWh = 3600 kJ.

**Line-commutated inverter**

An inverter that is tied into a power grid or line. The commutation of power (conversion from DC to AC) is controlled by the power line, so that if there is a failure in the power network, the PV system cannot feed power into the line.

**Load**

Anything in an electrical circuit which, when the circuit is turned on, draws power from that circuit.

**Maximum power point (MPP)**

The point on a current-voltage (IV) curve where maximum power is produced. For a typical silicon cell this is at about 0.45 V.

**Maximum power point tracker (MPPT)** — Means of a power conditioning unit that automatically operates the PV-generator at ist MPP under all conditions.

**Module** — See 'Photovoltaic module'.

**Mullion** — Slender pier which forms the division between the lights of a window, a screen or an opening.

**Mullion/Transom** — A popular facade construction often used with curtain wall (q.v.) facades. It consists of vertical beams (mullions) and smaller, horizontal beams (transoms).

**Multicrystalline silicon** — Silicon that has solidified at such a rate that many small crystals (crystallites) were formed. The atoms within a single crystallite are symmetrically arranged, whereas the crystallites are jumbled together. 'Multi' is used interchangeably with the prefix 'poly'.

**Muntin** — Upright piece of timber in a frame, separating panels. cf. 'Mullion'.

**Ohm** — The unit of resistance to the flow of an electric current.

**Open-circuit voltage ($V_{OC}$)** — The voltage across an illuminated photovoltaic cell or module when there is no current flowing; the $V_{OC}$ is the maximum possible voltage.

**Overhang** — Projection of a storey or any part of the building beyond a storey below or in front of the naked wall.

**Parallel connection** — A method of interconnecting two or more electricity-producing, or power-using devices, so that the voltage produced, or required, is not increased, but the current is additive. Opposite of Series connection (q.v.).

**Parapet** — Low wall to protect any place where there is a drop, as at the edge of a roof, balcony, terrace etc.

**Panel** — See 'Photovoltaic panel'.

**Peak watts ($W_p$)** — The amount of power a photovoltaic cell or module produces at nominal irradiation conditions (STC).

**Photovoltaic (PV) array** An interconnected system of photovoltaic panels that functions as a single electricity-producing unit. The panels are assembled as a discrete structure, with common support or mounting. In smaller systems, an array can consist of a single panel plus support structure or mounting.

**Photovoltaic (PV) cell** A photovoltaic cell is the smallest semi-conductor element within a PV module to perform the immediate conversion of light into electrical energy (DC voltage and current).

**Photovoltaic (PV) generator** A PV generator is the total of all PV strings of a PV power supply system, which are electrically interconnected.

**Photovoltaic (PV) module** The term 'module' is often used interchangeably with the term 'panel'.

**Photovoltaic (PV) panel** A group of modules fastened together and wired in either series or parallel. The term 'panel' is often used interchangeably with the term 'module'.

**Photovoltaic (PV) string** A PV string is a series connection of individual modules or equal groups of several paralleled modules.

**Photovoltaic (PV) system** A complete set of components for converting sunlight into electricity by the photovoltaic process, including array and balance of system components.

**Polycrystalline silicon** See 'Multicrystalline silicon'.

**Power density** The ratio of the power available from a battery to its mass (W/kg) or volume (W/l).

**Power conditioning equipment** Electrical equipment used to convert power from a photovoltaic array into a form suitable for subsequent use. A collective term for inverter, converter, battery charge regulator and blocking diode.

**Remote** Here: not connected to a utility grid.

**Semiconductor** A substance with conducting properties intermediate between those of a conductor and an insulator. In contrast to conductors, the resistance of a semicon-

| | |
|---|---|
| | ductor decreases with increasing temperature; other energizing forces, such as light, may also have the same effect. |
| **Series regulator** | Type of a battery charge regulator where the charging current is controlled by a switch connected in series with the PV-generator. |
| **Series connection** | A method of interconnecting devices that generate or use electricity so that the voltage, but not the current, is additive. Opposite of Parallel connection (q.v.). |
| **Shed** | Kind of flat roof skylight. |
| **Shelf life** | The length of time under specified conditions that a battery can be stored so that it keeps its guaranteed capacity. |
| **Short circuit current ($I_{SC}$)** | The current flowing freely from an illuminated photovoltaic cell or module through an external circuit that has no resistance. The $I_{SC}$ is the maximum current possible. |
| **Shunt regulator** | Type of a battery charge regulator where the charging current is controlled by a switch connected in parallel with the PV-generator. Overcharging of the battery is prevented by shorting the PV-generator. |
| **Skylight** | Frame containing glass or translucent/transparent material, set in a roof, fixed or opening. |
| **SOC** | 'State-of-Charge', the available capacity remaining in the battery expressed as a percentage of the rated capacity. |
| **Solar cell** | Same as photovoltaic cell. |
| **Standard Test Conditions (STC)** | Solar irradiation: 1000 W/m$^2$<br>Cell temperature: 25 °C<br>Spectrum: AM 1.5 |
| **Stand-alone (PV system)** | An autonomous or hybrid photovoltaic system not connected to a grid. May or may not have storage, but most stand-alone systems require batteries or some other form of storage. |

| | |
|---|---|
| **Stand-off mounting** | Technique for mounting a PV array on a sloped roof that involves mounting the modules a short distance above the pitched roof and tilting them to the optimum angle. |
| **Structural glazing** | A system of retaining glass or other materials to the aluminium members of a curtain wall using silicon sealant. These systems use no mechanical fasteners, and as a result have no profiles which cast shadows on the glazing surface. |
| **Transom** | Horizontal bar dividing a window or opening into two or more lights in height. |
| **Trickle charge** | A charge at a low rate, balancing losses through self-discharge to maintain a cell or battery in a fully charged condition. |
| **Truss** | Combination of timbers to form a frame, placed at intervals, carrying the purlins. As well as a frame, of timber or metal, the term means a projection from the face of a wall, or a large console. |
| **VAC** | Volts AC |
| **VDC** | Volts DC |
| **Volt (V)** | The unit of voltage which is a measure of the force or 'push' given the electrons in an electric circuit. One volt produces one ampere of current when acting against a resistance of one ohm. |
| **Watt (W)** | The unit of electric power or amount of work (J) done in a unit of time. One ampere of current flowing at a potential of one volt produces one watt of power. |
| **Watt-hour (Wh)** | see 'Kilowatt-hour' |

# Appendix IX

## GENERAL INFORMATION ABOUT THE IEA

INTERNATIONAL ENERGY AGENCY
The International Energy Agency, founded in November 1974, is an autonomous body within the framework of the Organization for Economic Cooperation and Development (OECD) which carries out a comprehensive program of energy cooperation among its 23 member countries. The European Commission also participates in the work of the Agency.

The policy goals of the IEA include diversity, efficiency and flexibility within the energy sector, the ability to respond promptly and flexible to energy emergencies, the environmentally sustainable provision and use of energy, more environmentally-acceptable energy sources, improved energy efficiency, research, development and market deployment of new and improved energy technologies, and cooperation among all energy market participants.

These goals are addressed in part through a program of international collaboration in the research, development and demonstration of new energy technologies under the framework of over 40 Implementing Agreements. The IEA's R&D activities are headed by the Committee on Energy Research and Technology (CERT) which is supported by a small Secretariat staff in Paris. In addition, four Working Parties (in Conservation, Fossil Fuels, Renewable Energy and Fusion) are charged with monitoring the various collaborative agreements, identifying new areas for cooperation and advising the CERT on policy matters.

IEA SOLAR HEATING AND COOLING PROGRAM
The Solar Heating and Cooling Program was one of the first collaborative R&D agreements to be established within the IEA, and, since 1977, its Participants have been conducting a variety of joint projects in active solar, passive solar and photovoltaic technologies, primarily for building applications. The twenty members are:

| | | |
|---|---|---|
| Australia | France | Spain |
| Austria | Germany | Sweden |
| Belgium | Italy | Switzerland |
| Canada | Japan | Turkey |
| Denmark | Netherlands | United Kingdom |
| European Commission | New Zealand | United States |
| Finland | Norway | |

A total of twenty-one projects or "Tasks" have been undertaken since the beginning of the Solar Heating and Cooling Program. The overall program is monitored by an Executive Committee consisting of one representative from each of the member countries. The leadership and management of individual Tasks are the responsibility of Operating Agents. These Tasks and their respective Operating Agents are:

| | |
|---|---|
| *Task 1: | Investigation of the Performance of Solar Heating and Cooling Systems - Denmark |
| *Task 2: | Coordination of Research and Development on Solar Heating and Cooling - Japan |
| *Task 3: | Performance Testing of Solar Collectors - Germany/United Kingdom |
| *Task 4: | Development of an Insulation Handbook and Instrument Package - United States |
| *Task 5: | Use of Existing Meteorological Information for Solar Energy Application - Sweden |
| *Task 6: | Solar Systems Using Evacuated Collectors - United States |
| *Task 7: | Central Solar Heating Plants with Seasonal Storage - Sweden |
| *Task 8: | Passive and Hybrid Solar Low Energy Buildings - United States |
| *Task 9: | Solar Radiation and Pyranometry Studies - Canada/Germany |
| *Task 10: | Solar Material Research and Testing - Japan |
| *Task 11: | Passive and Hybrid Solar Commercial Buildings - Switzerland |
| *Task 12: | Building Energy Analysis and Design Tools for Solar Applications - United States |
| Task 13: | Advanced Solar Low Energy Buildings - Norway |
| Task 14: | Advanced Active Solar Systems - Canada |
| Task 15: | Not initiated |
| Task 16: | Photovoltaics in Buildings - Germany |
| *Task 17: | Measuring and Modelling Spectral Radiation - Germany |
| Task 18: | Advanced Glazed Materials - United Kingdom |
| Task 19: | Solar Air Systems - Switzerland |
| Task 20: | Solar Energy in Building Renovation - Sweden |
| Task 21: | Daylighting in Buildings - Denmark |

*Completed

# Appendix X

## Participating Countries of the IEA Solar Heating and Cooling Programme Task 16 'Photovoltaics in Buildings'

Germany (Lead Country, Operating Agent Task 16: Dr. Heribert Schmidt)

| | | |
|---|---|---|
| Austria | Japan | Sweden |
| Canada | The Netherlands | Switzerland |
| Finland | Norway | United Kingdom |
| Spain | United States | |

Italy (Additional Contributor)

### Participants and Affiliations


| | | | |
|---|---|---|---|
| **Austria** | **Heinrich Wilk**<br>OKA<br>Böhmerwaldstr. 3<br>A - 4020 Linz | **Reinhard Haas**<br>TU Wien<br>Inst. f. Energiewirtschaft<br>Gusshausstr. 27-29/357<br>A - 1040 Wien | |
| **Canada** | **Ron LaPlace**<br>Photron Canada Inc.<br>P.O. Box 136<br>Colinton, Alberta T0G 0R0 | **Jimmy Royer**<br>SOLENER Inc.<br>442, rue Lavigueur<br>Quebec, Quebec G1R 1B5 | **V.S. Donepudi**<br>ESTCO Energy, Inc.<br>21 Concourse Gate - Unit 12<br>Nepean - Ontario K2E 7S4 |
| **Finland** | **Jyrki Leppänen**<br>Neste Oy<br>Corporate Technology<br>P.O. Box 310<br>FIN - 06101 Porvoo | **Peter D. Lund**<br>**Kimmo Peippo**<br>Helsinki University of<br>Technology - Department of<br>Technical Physics<br>Rakentajanaukio 2 C<br>FIN-02150 Espoo | **Asko Rasinkoski**<br>Neste Oy NAPS<br>Rälssitie 7<br>FIN - 01510 Vantaa |
| **Germany** | **Hermann Laukamp**<br>**Heribert Schmidt**<br>**Friedrich Sick**<br>Fraunhofer-Institut für<br>Solare Energiesysteme<br>Oltmannsstr. 5<br>D - 79100 Freiburg | **Thomas Erge**<br>Fraunhofer-Institut für<br>Solare Energiesysteme<br>Gruppe Leipzig<br>Zschortauer Str. 1A<br>D - 04129 Leipzig | **Heinz Hullmann**<br>Institut für Industrialisierung<br>des Bauens GmbH<br>Postfach 21 05 60<br>D - 30405 Hannover |
| | **Oussama Chehab**<br>Flachglas<br>Solartechnik GmbH<br>Muehlengasse 7<br>D - 50667 Köln | **Christian Bendel**<br>**Jürgen Schmid**<br>ISET<br>Königstor 59<br>D - 34119 Kassel | **Ingo Hagemann**<br>Planungsbüro Hagemann<br>Johanniterstr. 21<br>D - 52064 Aachen |
| | **Hans Gochermann**<br>DASA<br>Industriestr. 29<br>D - 22880 Wedel | | |

| | | | |
|---|---|---|---|
| **Italy (Additional Contributor)** | **Cinzia Abbate-Gardner**<br>Consultant to ENEL<br>P.za S. Anastasia, 3<br>I - 00186 Roma | | |
| **Japan** | **Shogo Nishikawa**<br>Technology & Development Laboratory, Research & Development Headquarters, KANDENKO Co., Ltd.<br>2673-169 Shimoinayoshi, Nishiyama Chiyodamachi, Niihari-gun, Ibaraki-ken 315 | | |
| **Netherlands** | **Tony Schoen**<br>ECOFYS<br>P.O. Box 8408<br>NL - 3503 RK Utrecht | **Karel van Otterdijk**<br>E C N<br>Westerduinweg 3<br>P.O. Box 1<br>NL - 1755 ZG Petten (N.H.) | **Emil W. Ter Horst**<br>Novem<br>P.O. Box 8242<br>NL - 3505 RE Utrecht |
| **Norway** | **Inger Andresen**<br>SINTEF Architecture and Building Technology<br>Alfred Getz Vei 3<br>N - 7034 Trondheim | **Oyvin Skarstein**<br>Norwegian Institute of Technology, Dept. of Electric Power Engineering<br>N - 7034 Trondheim | |
| **Spain** | **Alvaro Gonzalez Menendez**<br>Ciemat-IER<br>Avda. Complutense 22<br>Ed. 42<br>E - 28040 Madrid | | |
| **Sweden** | **Mats Andersson**<br>Catella Generics AB<br>Bäling<br>S - 82077 Gnarp | **Bengt Perers**<br>Vattenfall Utveckling AB<br>c/o Miljö konsulterna<br>Box 154<br>S - 61124 Nyköping | |
| **Switzerland** | **Peter Toggweiler**<br>PMS Energie AG<br>Lindhofstr. 13<br>CH - 8617 Mönchaltorf | **Christian Roecker**<br>EPFL-LESO<br>Bâtiment LESO<br>CH - 1015 Lausanne | **Christophe De Reyff**<br>Swiss Fed. Office of Energy<br>Belpstr. 36<br>CH - 3003 Bern |
| **United Kingdom** | **Frances Crick**<br>**Jean-Paul Louineau**<br>**Bernard McNelis**<br>IT Power Ltd.<br>The Warren, Bramshill Road<br>Eversley, Hants. RG27 0PR | **R.D.W. Scott**<br>BP Solar<br>PO Box 191, Chertsey Road<br>Sunbury-on-Thames<br>Middx. TW16 7XA | **Harry Edwards**<br>ETSU<br>for the Department of Trade & Industry<br>Harwell, Didcot<br>Oxfordshire OX11 ORA |
| **United States** | **Steven J. Strong**<br>Solar Design Associates Inc.<br>P.O. Box 242<br>Harvard, MA 01451-0242 | **Robert J. Hassett**<br>U.S. Dept. of Energy<br>MS 5H048<br>1000 Independence Ave., S.W.,<br>Washington, DC 20585 | **Sheila J. Hayter**<br>**Roger W. Taylor**<br>National Renewable Energy Laboratory<br>1617 Cole Boulevard<br>Golden, CO 80401 |

# Index